The Enigmatic Photon

Fundamental Theories of Physics

An International Book Series on The Fundamental Theories of Physics: Their Clarification, Development and Application

Volume 106

The Enigmatic Photon

Volume 5: O(3) Electrodynamics

by

Myron W. Evans

Alpha Foundation and Laboratories,
Institute of Physics,
Budapest, Hungary

KLUWER ACADEMIC PUBLISHERS

DORDRECHT / BOSTON / LONDON

A C.I.P. Catalogue record for this book is available from the Library of Congress.

ISBN 0-7923-5792-2

Published by Kluwer Academic Publishers,
P.O. Box 17, 3300 AA Dordrecht, The Netherlands.

Sold and distributed in North, Central and South America
by Kluwer Academic Publishers,
101 Philip Drive, Norwell, MA 02061, U.S.A.

In all other countries, sold and distributed
by Kluwer Academic Publishers,
P.O. Box 322, 3300 AH Dordrecht, The Netherlands.

Printed on acid-free paper

Printed in the Netherlands.

Contents

Preface

The first part of this fifth volume of *The Enigmatic Photon* consists of three chapters which develop electrodynamics as a gauge field theory with internal *O(3)* symmetry using a complex basis ((1), (2), (3)). This is referred to for convenience as *O(3)* electrodynamics. The field equations are intrinsically non-linear and non-Abelian, and the classical potential is a physical quantity. The field equations are shown to be self-consistent and a general mathematical basis for the development is given in terms of extended Lie algebra. The third chapter develops radiatively induced fermion resonance (*RFR*).

The opening three chapters are followed by a selection of previously unpublished scientific papers which trace the historical development of *O(3)* electrodynamics through the physically observable $\boldsymbol{B}^{(3)}$ field.

The editors are greatly indebted to Dr. Laura J. Evans for her painstaking preparation of Part I and Part II of this volume. This work included typesetting from original manuscripts and indexing. The production of the five volumes of this series would not have been possible without her steadfast and generous contributions.

Discussions with many colleagues over internet sharpened the ideas behind this volume. Finally, the organizational support of the Alpha Foundation, founded by Milán Mészáros, proved invaluable during the later preparatory stages.

Ithaca, NY, USA Myron W. Evans
October, 1998

Part I

Electrodynamics as a Gauge Field Theory

Chapter 1

General Gauge Field Theory Applied to Electrodynamics

In the first part of this fifth volume it is argued that electrodynamics can be developed self consistently as an example of contemporary general gauge field theory. The basic assumption in this development is that the left and right circular polarization discovered by Arago in 1811 can be supplemented by a longitudinal component, (3), forming a complex circular basis ((1), (2), (3)) of *O(3)* symmetry - the symmetry of the rotation group. The nature of the basis has been elaborated in Vols. 1 to 4 [1—4] of this series so we take advantage of this groundwork in this volume to try to establish the complete self consistency of *O(3)* electrodynamics on the classical level. This means that the fields are described classically in terms of physical potentials within general gauge field theory [5]. The latter borrows concepts from general relativity, the most important of which is the covariant derivative [1—5], used extensively in the first four volumes. In *O(3)* symmetry, the field tensor in classical electrodynamics is made up of terms both linear and non-linear in the potential, which is a vector in an internal gauge space ((1), (2), (3)). This space is superimposed on space-time in such a way that indices are matched self consistently, forming an extended Lie algebra in which the spaces are not independent.

The rules of general gauge field theory, rules that have led to the discovery of quarks, for example, are then applied to electrodynamics in this *O(3)* group symmetry and several results obtained which are absent from, or ill defined in, the received *U(1)* electrodynamics. Under certain well defined conditions, non linear *O(3)* electrodynamics are well approximated by the linear *U(1)* electrodynamics. For the free field, however, the *O(3)* gauge field symmetry leads to a novel, always non-zero, fundamental field $\boldsymbol{B}^{(3)}$ [1—4], and to novel concepts such as classical vacuum polarization and magnetization which are missing from *U(1)* electrodynamics entirely. The vector potential in *O(3)* symmetry is a classical object, and the rules of gauge transformation are different from those in the older view. This conclusion leads to many ramifications and concepts which are also missing from *U(1)*. *These concepts are due to the non Abelian nature of the O(3)* theory, for example inherent non linearities such as the well observed conjugate product $\boldsymbol{A}^{(1)} \times \boldsymbol{A}^{(2)}$ of complex vector potentials discussed throughout Vols. 1—4 and elsewhere [5] in the literature.

In the presence of field matter interaction it is shown that the *U(1)* theory can be recovered as an excellent approximation to the *O(3)* theory, because when there is field matter interaction the non linear terms are very small, empirically and theoretically. This correct recovery of the linear *U(1)* field equations from those of the non-linear *O(3)* theory means that the latter can do everything that the former can do plus a lot more. The Coulomb, Ampère, Faraday and Gauss laws can be recovered from the *O(3)* theory when the latter's non linearities can be neglected.

For the free field, however, the non-linearities of the *O(3)* theory are intrinsically important and cannot be approximated or gauged away. For example, the $\boldsymbol{B}^{(3)}$ field of the *O(3)* theory does not exist in the *U(1)* theory. The inverse Faraday effect can be accounted for from first principles in the *O(3)* theory, but it leads to a paradox in the *U(1)* theory. In the latter, the potentials can be regarded as mathematical subsidiary variables [7—9], but in the *O(3)* theory they are physically meaningful, for example, there is a light like $\left(cA^{(0)}, \boldsymbol{A}^{(3)}\right)$, a polar four-vector that quantizes directly to photon momentum and which is missing entirely from the *U(1)* theory. Gauge transformation in the *O(3)* theory is a geometrical process with physical meaning, whereas in the *U(1)* theory it is essentially a mathematical process using the gradient of an arbitrary variable. One consequence is that in the *O(3)* theory the Lorentz transformation has a different meaning; if it applies at all it is a special restriction on the physical vector potential. In the *U(1)* theory it is a key choice of gauge that is ultimately a mathematical convenience, leading as it does to the d'Alembert wave equation and to the gauge fixing term used in *U(1)* quantization.

The empirical evidence for the need for an *O(3)* or *SU(2)* symmetry for classical electrodynamics has been reviewed recently by Barrett [8,9], who argues that the classical Maxwellian view of electrodynamics is a linear theory in which the scalar and vector potentials are arbitrary, and defined only through applied boundary conditions and a subjective choice of gauge such as the Lorentz condition. Barrett [8,9] then exposes several flaws in the received view by arguing that there exist several phenomena of nature that require a physical potential four-vector on the classical as well as the quantum levels. One of these is the Aharonov-Bohm effect, but there are several others. The examples thus far identified include the following: 1) Aharonov-Bohm; 2) Altshuler-Aronov-Spivak; 3) topological phase; 4) Josephson; 5) quantized Hall; 6) Sagnac; 7) de Haas van Alphen; 8) Ehrenberg-Siday; 9) non-linear magneto optical. It is also significant that quantum electrodynamics leads to vacuum polarization, or photon self energy, which is missing from classical *U(1)* theory but is present in classical *O(3)* theory as shown in this chapter. The *O(3)* theory also gives classical vacuum magnetization, also missing from *U(1)* theory.

So to accept the suggestion of *O(3)* electrodynamics it is necessary to consider the empirical data given by Barrett, and to accept the hypothesis that gauge field theory can be developed with *O(3)* covariant derivatives which can be classified with group theory and which can be applied to classical electrodynamics [1—4]. Once this hypothesis is accepted and tested for self-consistency, several advantages follow which are described in the first part of this volume. Resistance to the hypothesis based on the standard model is counter-argued in Refs.1 through 4 and on the key empirical observations listed above, for example those by Barrett [8,9] and the empirical observation [1—4] of the conjugate product $\boldsymbol{A}^{(1)} \times \boldsymbol{A}^{(2)}$. This is rigorously zero in *U(1)* electrodynamics by definition [1—6], but is non-zero in *O(3)* electrodynamics. There is no difficulty in principle in extending quantum electrodynamics to a non-Abelian theory, which becomes akin to quantum chromodynamics. The latter is well known to be renormalizable at all orders. The mathematical structure of non Abelian *qed* is that of *qcd*, but with an internal gauge space ((1), (2), (3)). As shown in Chap. 2, the gauge space and space-time form an extended Lie algebra in electrodynamics, even in the *U(1)* theory. The two spaces are not independent of each other, even in the standard model.

Therefore the first part of this volume develops the ideas of *O(3)* electrodynamics, giving an unusual amount of technical detail because the hypothesis and concomitant ideas may be new to the classical electrodynamicist versed in the standard model, which allows only *U(1)* theory for the electromagnetic sector. The contemporary gauge field theorist will be unfamiliar with the fact that the internal gauge space and the space-time of both the *U(1)* and the *O(3)* theory are not independent (Chap. 2), in the sense that they form an extended Lie algebra as discussed elegantly by Aldrovandi [10]. The concept of *O(3)* electrodynamics must not be confused as an abstract analogy of the *U(1)* electrodynamics. The former produces physical equations of classical electrodynamics which reduce to the form of the Maxwell equations for polarizations (1) and (2), so in this sense the *O(3)* (non-linear) theory, reduces to the *U(1)* theory when non linearities are small. This occurs in field matter interaction. For example, the non-linear inverse Faraday effect is in magnitude a very small effect of magneto optics which was finally observed with considerable difficulty in 1965 [1—4]. The linear Maxwell equations describe the much more accessible and more easily observable optical effects of nature that go back to the discovery of circular polarization by Arago in 1811, and to the work of Coulomb in the late eighteenth century. The Maxwell equations work well because the optical non linearities in nature are so small in field matter interaction.

At the risk of boring the initiated therefore, we provide throughout the opening chapters of this volume copious details of the new theory, to try to minimize confusion and obscurity, and to help the student. The first section of this chapter deals with the fundamental vector algebra of the complex circular basis ((1), (2), (3)), showing that it is, indeed, a basis that can be used as a representation of *O(3)* space. As intimated, the use of this basis is suggested by the empirical existence of right and left circular polarization, which must be described in a complex representation by at least two basis vectors, $\boldsymbol{i}$ and $\boldsymbol{j}$ in the Cartesian representation, $\boldsymbol{e}^{(1)}$ and $\boldsymbol{e}^{(2)}$ in the complex circular representation. As argued elegantly by Barrett [8,9] this basic fact about light leads to the need for an *SU(2)* electrodynamics.

In our view, the $\boldsymbol{B}^{(3)}$ field emerges once we accept an *SU(2)* or *O(3)* electrodynamics for the vacuum as well as for field matter interaction. It turns out that the hypothesis of *O(3)* electrodynamics is self consistent and is as valid in this sense as *U(1)* electrodynamics. It is recognized however that all Maxwellian type theories have serious flaws inherent in them, and the extension from *U(1)* to *O(3)* does not cure all of these. The standard model is rigidly cemented in *U(1)* theory and carries with it all these serious flaws listed, for example by Bearden [11], and recently discussed by Fritzius' translation [12] of Ritz [13]. These have each argued elegantly against the *U(1)* electrodynamics for a number of years.

1.1 Elements of Vector Analysis in the Circular Basis ((1), (2), (3))

The ((1), (2), (3)) basis is hereinafter referred to as the complex circular basis because it is formed from a complex combination of Cartesian unit vectors as they appear in the description of circular polarization. The basis vectors are therefore,

$$\boldsymbol{e}^{(1)} = \frac{1}{\sqrt{2}}\left(\boldsymbol{i} - i\boldsymbol{j}\right), \qquad \boldsymbol{i} = \frac{1}{\sqrt{2}}\left(\boldsymbol{e}^{(1)} + \boldsymbol{e}^{(2)}\right),$$

$$\boldsymbol{e}^{(2)} = \frac{1}{\sqrt{2}}\left(\boldsymbol{i} + i\boldsymbol{j}\right), \qquad \boldsymbol{j} = \frac{i}{\sqrt{2}}\left(\boldsymbol{e}^{(1)} - \boldsymbol{e}^{(2)}\right), \tag{1.1.1}$$

$$\boldsymbol{e}^{(3)} = \boldsymbol{k}\,.$$

If the phase factor $e^{-i\phi}$ of electromagnetic radiation is kept constant, then $\boldsymbol{e}^{(1)} = \boldsymbol{e}^{(2)*}$ is the vectorial part of the circular description of right and left circularly polarized radiation. Note carefully however that in forming the complex conjugate of a plane wave such as $\boldsymbol{B}^{(1)}$ the phase factor also changes from $e^{-i\phi}$ to $e^{i\phi}$. These matters are described at length in Ref. 14.

The vectors $\boldsymbol{e}^{(1)}, \boldsymbol{e}^{(2)}$ and $\boldsymbol{e}^{(3)}$ form the *O(3)* type cyclic permutation relations [1—4],

$$\begin{aligned} \boldsymbol{e}^{(1)}\times\boldsymbol{e}^{(2)} &= i\boldsymbol{e}^{(3)*}, & \boldsymbol{i}\times\boldsymbol{j} &= \boldsymbol{k}, \\ \boldsymbol{e}^{(2)}\times\boldsymbol{e}^{(3)} &= i\boldsymbol{e}^{(1)*}, & \boldsymbol{j}\times\boldsymbol{k} &= \boldsymbol{i}, \\ \boldsymbol{e}^{(3)}\times\boldsymbol{e}^{(1)} &= i\boldsymbol{e}^{(2)*}, & \boldsymbol{k}\times\boldsymbol{i} &= \boldsymbol{j}. \end{aligned} \tag{1.1.2}$$

A closely similar complex circular basis has been described for example by Silver [15], and is well known in tensor analysis.

1.1.1 The Unit Vector Dot Product

In the complex circular basis,

$$\begin{aligned} \boldsymbol{e}^{(1)}\cdot\boldsymbol{e}^{(2)} = \boldsymbol{e}^{(2)}\cdot\boldsymbol{e}^{(1)} = \boldsymbol{e}^{(3)}\cdot\boldsymbol{e}^{(3)} = 1 \\ \boldsymbol{e}^{(1)}\cdot\boldsymbol{e}^{(1)} = \boldsymbol{e}^{(2)}\cdot\boldsymbol{e}^{(2)} = 0\ . \end{aligned} \tag{1.1.3}$$

In the Cartesian basis,

$$\boldsymbol{i}\cdot\boldsymbol{i} = \boldsymbol{j}\cdot\boldsymbol{j} = \boldsymbol{k}\cdot\boldsymbol{k} = 1\ , \qquad \boldsymbol{i}\cdot\boldsymbol{j} = \boldsymbol{i}\cdot\boldsymbol{k} = \boldsymbol{j}\cdot\boldsymbol{k} = \boldsymbol{0}\ . \tag{1.1.4}$$

1.1.2 Vectors

In the complex circular basis the vectors $\boldsymbol{A}$ and $\boldsymbol{B}$ can be defined as

$$\begin{aligned} \boldsymbol{A} &:= \boldsymbol{A}^{(1)} + \boldsymbol{A}^{(2)} + \boldsymbol{A}^{(3)} = A^{(1)}\boldsymbol{e}^{(1)} + A^{(2)}\boldsymbol{e}^{(2)} + A^{(3)}\boldsymbol{e}^{(3)}, \\ \boldsymbol{B} &:= \boldsymbol{B}^{(1)} + \boldsymbol{B}^{(2)} + \boldsymbol{B}^{(3)} = B^{(1)}\boldsymbol{e}^{(1)} + B^{(2)}\boldsymbol{e}^{(2)} + B^{(3)}\boldsymbol{e}^{(3)}, \end{aligned} \tag{1.1.5}$$

In these definitions, $A^{(1)}$, $A^{(2)}$ and $A^{(3)}$ are scalars, linked to their Cartesian counterparts as follows,

$$A^{(1)} = \frac{1}{\sqrt{2}}\left(A_X - iA_Y\right) = A^{(2)*} , \qquad A^{(3)} = A_Z . \tag{1.1.6}$$

1.1.2.1 Unit Vector Premultipliers

In the logic of the complex circular basis scalar unity is expressed as the product of two complex conjugates, referred to here as *complex unity*,

$$1^2 := 1^{(1)}1^{(2)} , \tag{1.1.7}$$

where,

$$1^{(1)} = \frac{1}{\sqrt{1}}\left(1 - i\right) , \qquad 1^{(2)} = \frac{1}{\sqrt{2}}\left(1 + i\right) , \tag{1.1.8}$$

so the dot product of $\boldsymbol{e}^{(1)}$ with $\boldsymbol{e}^{(2)}$ or of vectors $\boldsymbol{A}^{(1)}$ and $\boldsymbol{A}^{(2)}$ is

$$\left.\begin{aligned} \boldsymbol{e}^{(1)} \cdot \boldsymbol{e}^{(2)} &= 1^{(1)}\boldsymbol{e}^{(1)} \cdot 1^{(2)}\boldsymbol{e}^{(2)} = 1^{(1)}1^{(2)} = 1^2 = 1 , \\ \boldsymbol{A}^{(1)} \cdot \boldsymbol{A}^{(2)} &= A^{(1)}\boldsymbol{e}^{(1)} \cdot A^{(2)}\boldsymbol{e}^{(2)} = A^{(1)}A^{(2)} = A^{(0)2} . \end{aligned}\right\} \tag{1.1.9}$$

Since the product $1^{(1)}1^{(2)}$ is always unity, it makes no difference to the dot product of unit vectors or of conjugate vectors such as $\boldsymbol{A}^{(1)}$ and $\boldsymbol{A}^{(2)}$, but the dot product of a vector $\boldsymbol{A}^{(1)}$ and a unit vector $\boldsymbol{e}^{(2)}$ is

$$\begin{aligned} \boldsymbol{A}^{(1)} \cdot \boldsymbol{e}^{(2)} &= A^{(1)}1^{(2)}\boldsymbol{e}^{(1)} \cdot \boldsymbol{e}^{(2)} = \frac{1}{2}\left(A_X - iA_Y\right)\left(1 + i\right) \\ &= \frac{1}{2}\left(A_X - iA_Y + iA_X + A_Y\right) . \end{aligned} \tag{1.1.10}$$

Similarly, as described in Appendix B of Vol. 3, the dot product of a complex circular Pauli matrix $\sigma^{(1)}$ and a unit vector $\boldsymbol{e}^{(2)}$ is

$$\sigma^{(1)} \cdot \boldsymbol{e}^{(2)} = \frac{1}{2}\left(\sigma_X - i\sigma_Y + i\sigma_X + \sigma_Y\right), \tag{1.1.11}$$

as in that Appendix. This procedure leads to the result of that appendix,

$$\left(\sigma^{(1)} \cdot \boldsymbol{e}^{(2)}\right)\left(\sigma^{(2)} \cdot \boldsymbol{e}^{(2)}\right)^{+} = \boldsymbol{e}^{(1)} \cdot \boldsymbol{e}^{(2)} + i\sigma^{(3)} \cdot \boldsymbol{e}^{(1)} \times \boldsymbol{e}^{(2)}, \tag{1.1.12}$$

which is the equivalent of the Dirac result of the Cartesian basis.

The complex circular basis is a natural description of the observable conjugate product $\boldsymbol{A}^{(1)} \times \boldsymbol{A}^{(2)}$ and thus of $\boldsymbol{B}^{(3)}$.

1.1.3 Dot Product of Two Vectors

The dot product of two vectors when neither is a unit vector is defined as

$$\begin{aligned}\boldsymbol{A} \cdot \boldsymbol{B} &= A^{(1)}B^{(2)}\boldsymbol{e}^{(1)} \cdot \boldsymbol{e}^{(2)} + A^{(2)}B^{(1)}\boldsymbol{e}^{(2)} \cdot \boldsymbol{e}^{(1)} + A^{(3)}B^{(3)}\boldsymbol{e}^{(3)} \cdot \boldsymbol{e}^{(3)} \\ &= A^{(1)}B^{(2)} + A^{(2)}B^{(1)} + A^{(3)}B^{(3)},\end{aligned} \tag{1.1.13}$$

and is the same as the Cartesian dot product,

$$\boldsymbol{A} \cdot \boldsymbol{B} = A_X B_X + A_Y B_Y + A_Z A_Z . \tag{1.1.14}$$

1.1.4 The Del Operator

The del operator in the complex circular basis is a vector operator which can be defined as

$$\nabla_X = \frac{\partial}{\partial X} = \frac{1}{\sqrt{2}}\left(\nabla^{(1)} + \nabla^{(2)}\right), \qquad \nabla^{(1)} = \frac{1}{\sqrt{2}}\left(\nabla_X - i\nabla_Y\right),$$

$$\nabla_Y = \frac{\partial}{\partial Y} = \frac{i}{\sqrt{2}}\left(\nabla^{(1)} - \nabla^{(2)}\right), \qquad \nabla^{(2)} = \frac{1}{\sqrt{2}}\left(\nabla_X + i\nabla_Y\right), \tag{1.1.15}$$

$$\nabla_Z = \frac{\partial}{\partial Z} = \nabla^{(3)}, \qquad \nabla^{(3)} = \nabla_Z .$$

1.1.5 Divergence

The divergence in the complex circular basis is defined as

$$\nabla \cdot \boldsymbol{A} = \nabla^{(1)} A^{(2)} + \nabla^{(2)} A^{(1)} + \nabla^{(3)} A^{(3)} . \tag{1.1.16}$$

1.1.6 Gradient

The gradient of a scalar Φ in the complex circular basis is,

$$\nabla \Phi = \nabla^{(1)} \Phi \boldsymbol{e}^{(2)} + \nabla^{(2)} \Phi \boldsymbol{e}^{(1)} + \nabla^{(3)} \Phi \boldsymbol{e}^{(3)} . \tag{1.1.17}$$

1.1.7 Curl

The curl operator in the complex circular basis is defined as,

$$\nabla \times \boldsymbol{A} = \begin{vmatrix} \boldsymbol{i} & \boldsymbol{j} & \boldsymbol{k} \\ \nabla_X & \nabla_Y & \nabla_Z \\ A_X & A_Y & A_Z \end{vmatrix} = -i \begin{vmatrix} \boldsymbol{e}^{(1)} & \boldsymbol{e}^{(2)} & \boldsymbol{e}^{(3)} \\ \nabla^{(1)} & \nabla^{(2)} & \nabla^{(3)} \\ A^{(1)} & A^{(2)} & A^{(3)} \end{vmatrix} . \tag{1.1.18}$$

For example the $\boldsymbol{k}$ component is

$$\left(\nabla_X A_Y - \nabla_Y A_X\right)\boldsymbol{k} = -i\left(\nabla^{(1)} A^{(2)} - \nabla^{(2)} A^{(1)}\right)\boldsymbol{e}^{(3)} . \tag{1.1.19}$$

The $\boldsymbol{i}$ and $\boldsymbol{j}$ components are

$$\begin{aligned}
&\boldsymbol{i}\left(\nabla_Y A_Z - \nabla_Z A_Y\right) - \boldsymbol{j}\left(\nabla_X A_Z - \nabla_Z A_X\right) = \frac{1}{\sqrt{2}}\left(\boldsymbol{e}^{(1)} + \boldsymbol{e}^{(2)}\right) \\
&\times\left(\frac{i}{\sqrt{2}}\left(\nabla^{(1)} - \nabla^{(2)}\right) A^{(3)} - \frac{i}{\sqrt{2}}\nabla^{(3)}\left(A^{(1)} - A^{(2)}\right)\right) \\
&- \frac{i}{\sqrt{2}}\left(\boldsymbol{e}^{(1)} - \boldsymbol{e}^{(2)}\right)\left(\frac{1}{\sqrt{2}}\left(\nabla^{(1)} + \nabla^{(2)}\right) A^{(3)} - \frac{\nabla^{(3)}}{\sqrt{2}}\left(A^{(1)} + A^{(2)}\right)\right) \\
&= -i\left(\left(\nabla^{(2)} A^{(3)} - \nabla^{(3)} A^{(2)}\right)\boldsymbol{e}^{(1)} + \left(\nabla^{(3)} A^{(1)} - \nabla^{(1)} A^{(3)}\right)\boldsymbol{e}^{(2)}\right) .
\end{aligned} \tag{1.1.20}$$

1.1.8 The Vector Cross Product

The vector cross product in the complex circular basis is by definition,

$$\boldsymbol{A} \times \boldsymbol{B} := \left(A^{(2)}\boldsymbol{e}^{(1)} + A^{(1)}\boldsymbol{e}^{(2)} + A^{(3)}\boldsymbol{e}^{(3)}\right)$$

$$\times \left(B^{(2)}\boldsymbol{e}^{(1)} + B^{(1)}\boldsymbol{e}^{(2)} + B^{(3)}\boldsymbol{e}^{(3)}\right)$$

$$= A^{(2)}B^{(1)}\boldsymbol{e}^{(1)} \times \boldsymbol{e}^{((2)} + \ \ldots \ = i\boldsymbol{e}^{(3)*}A^{(2)}B^{(1)} + \ \ldots \tag{1.1.21}$$

$$= i\begin{vmatrix} \boldsymbol{e}^{(1)*} & \boldsymbol{e}^{(2)*} & \boldsymbol{e}^{(3)*} \\ A^{(2)} & A^{(1)} & A^{(3)} \\ B^{(2)} & B^{(1)} & B^{(3)} \end{vmatrix} .$$

This result can be checked by working out the $\boldsymbol{e}^{(3)}$ component of $\boldsymbol{A} \times \boldsymbol{B}$, $i\boldsymbol{e}^{(3)*}\left(A^{(2)}B^{(1)} - A^{(1)}B^{(2)}\right)$, where,

$$A^{(2)} = \frac{1}{\sqrt{2}}\left(A_X + iA_Y\right) = A^{(1)*} ,$$

$$B^{(1)} = \frac{1}{\sqrt{2}}\left(B_X - iB_Y\right) = B^{(2)*} . \tag{1.1.21a}$$

So,

$$A^{(2)}B^{(1)} - B^{(2)}A^{(1)} = -i\left(A_X B_Y - B_X A_Y\right)$$

$$\text{and } \left(A_X B_Y - A_Y B_X\right)\boldsymbol{k} = i\boldsymbol{e}^{(3)*}\left(A^{(2)}B^{(1)} - B^{(2)}A^{(1)}\right) . \tag{1.1.22}$$

The conjugate product can be checked by direct evaluation to be

$$\boldsymbol{A}^{(1)} \times \boldsymbol{A}^{(2)} = \begin{vmatrix} \boldsymbol{i} & \boldsymbol{j} & \boldsymbol{k} \\ A_X^{(1)} & A_Y^{(1)} & 0 \\ A_X^{(2)} & A_Y^{(2)} & 0 \end{vmatrix} \tag{1.1.23}$$

$$= \left(A_X^{(1)} A_Y^{(2)} - A_Y^{(1)} A_X^{(2)} \right) \boldsymbol{k} = iA^{(0)} \boldsymbol{A}^{(3)*} \, .$$

1.1.9 The Cyclic Relations

The cyclic relations in the ((1), (2), (3)) basis are

$$\boldsymbol{A}^{(1)} \times \boldsymbol{A}^{(2)} = iA^{(0)} \boldsymbol{A}^{(3)*} \, , \quad \text{et cyclicum} \tag{1.1.24}$$

and so on for any vector.

1.2 The Electromagnetic Field Tensor in *O(3)* Electrodynamics

The basic concepts of this section were first tried out in Vol. 2 but here we offer a considerable clarification and simplification based on intervening experience and discussion. The basic ideas of general field theory are described for example in Ryder [5] in his Chap. 3, and were first applied to electrodynamics in Vol. 2 in a didactic manner. In order to understand these ideas at all, two concepts in particular are needed which do not exist in *U(1)* electrodynamics: that of the internal space and the covariant derivative as defined in this space. These new and perhaps unfamiliar ideas are best illustrated when it comes to gauge transformation in *O(3)*, which is developed in the next section. This section gives basic ideas and at each stage is careful to spot the difference between *U(1)* and *O(3)*. In this way the interested student can gradually absorb the new material and realize its advantages. In so doing the need for an *O(3)* electrodynamics becomes ever clearer, and several advantages over *U(1)* start to be defined.

1.2.1 The Internal Space in *O(3)* Electrodynamics

The internal space is defined through the expansion of the potential in space-time to an object which has meaning additionally in a space defined by a particular group structure [1—6]. It becomes necessary to think of the familiar A^{μ} of the received view as a scalar object in this internal space as well as a four-vector in space-time. This idea appears to have been first applied to field theory by Yang and Mills in 1955 [8,9], and has since been developed in many very fruitful ways within the standard model. It was first applied by Barrett [8,9] to classical electrodynamics in the late eighties, and slightly later [1—4] it was shown to lead to the existence of the fundamental field $\boldsymbol{B}^{(3)}$, an object that is missing from the received view. It is not so much that the latter sets $\boldsymbol{B}^{(3)}$ to zero, it is a concept that simply does not appear within its horizon. So it is clear that *O(3)* or *SU(2)* electrodynamics was inferred independently by Barrett [8,9] and by Evans [1—4].

Therefore if A^{μ} is thought of as a scalar object in some internal space, conceptually and empirically ((1), (2), (3)), it becomes possible to write a potential that becomes a vector in the internal space, and whose scalar components in this space are also objects in space-time. This is the basic hypothesis of *O(3)* electrodynamics, and we can write in consequence of this hypothesis,

$$\boldsymbol{A}^{\mu} = A^{\mu(1)}\boldsymbol{e}^{(1)} + A^{\mu(2)}\boldsymbol{e}^{(2)} + A^{\mu(3)}\boldsymbol{e}^{(3)} . \qquad (1.1.25)$$

The unit vectors $\boldsymbol{e}^{(1)}$, $\boldsymbol{e}^{(2)}$, and $\boldsymbol{e}^{(3)}$ form a complex basis for internal space, and the objects $A^{\mu(1)}$, $A^{\mu(2)}$ and $A^{\mu(3)}$ are scalar coefficients in the internal space of the complete vector $\boldsymbol{A}^{\mu}$. This boldface character therefore denotes an object that is simultaneously a vector in the internal space (a symmetry space [8,9] of gauge field theory) and a four-vector in space-time

(Minkowski space). The indices of the scalar coefficients $A^{\mu(1)}$, $A^{\mu(2)}$ and $A^{\mu(3)}$ must therefore match self consistently.

1.2.1.1 Index Matching

If we consider the received view of ordinary plane waves in space-time [1—4],

$$\boldsymbol{A}^{(1)} = \boldsymbol{A}^{(2)*} = \frac{A^{(0)}}{\sqrt{2}} \left(i\boldsymbol{i} + \boldsymbol{j} \right) e^{-i\phi} , \tag{1.1.26}$$

it should be clear that the boldface character $\boldsymbol{A}^{(1)}$ represents a vector in the ordinary space part of space-time. The electromagnetic phase is defined as $\phi := \omega t - \kappa Z$ where ω is the angular frequency at instant t and κ the wave-vector at point Z.

These plane waves are transverse solutions of the received *U(1)* field equations and the d'Alembert wave equation for the free field [1—9]. In order to expand the horizon of the gauge structure of classical electrodynamics from *U(1)* to *O(3)* an additional space-time index must appear in the definition of the plane wave and the (1) and (2) indices must become indices of the internal space. This is achieved by recognizing that:

$$\left.\begin{aligned} A^{1(1)} &= A_X^{(1)} = i\frac{A^{(0)}}{\sqrt{2}} e^{-i\phi} = A^{1(2)*} , \\ A^{2(1)} &= A_Y^{(1)} = \frac{A^{(0)}}{\sqrt{2}} e^{-i\phi} = A^{2(2)*} , \\ A^{0(1)} &= A^{3(1)} = A^{0(2)} = A^{3(2)} = 0 , \end{aligned}\right\} \tag{1.1.27}$$

These equations define two of the scalar coefficients of the complete four-vector $\boldsymbol{A}^{\mu}$,

$$\left.\begin{aligned} A^{\mu(1)} &= \left(0, \boldsymbol{A}^{(1)}\right) \\ A^{\mu(2)} &= \left(0, \boldsymbol{A}^{(2)}\right) . \end{aligned}\right\} \qquad (1.1.28)$$

This process follows from the fact that $\boldsymbol{A}^{(1)} = \boldsymbol{A}^{(2)*}$ are transverse, and so can have *X* and *Y* components only. The scalar coefficients $A^{\mu(1)}$ and $A^{\mu(2)}$ are light-like invariants [16,17],

$$A^{\mu(1)} A_{\mu}^{(1)} = A^{\mu(2)} A_{\mu}^{(2)} = 0 , \qquad (1.1.29)$$

of polar four-vectors in space-time. The third index (3) of the non Abelian theory must therefore be along the direction of propagation of the radiation and must also be a light-like invariant,

$$A^{\mu(3)} A_{\mu}^{(3)} = 0 , \qquad (1.1.30)$$

in the vacuum.. It must be light-like because the free field is assumed to propagate, in this classical view, at c in the vacuum..

One possible solution of Eq. (1.1.30) is

$$A^{\mu(3)} = \left(cA^{(0)}, \boldsymbol{A}^{(3)}\right) , \qquad (1.1.31)$$

where

$$cA^{(0)} = \left|\boldsymbol{A}^{(3)}\right| . \qquad (1.1.32)$$

Such a solution is proportional directly to the wave four-vector,

$$\kappa^{\mu(3)} := \left(c\kappa, \kappa \boldsymbol{e}^{(3)}\right) = eA^{\mu(3)} , \qquad (1.1.33)$$

and to the photon energy-momentum,

$$p^{\mu(3)} := \hbar\kappa^{\mu(3)} = eA^{\mu(3)} , \qquad (1.1.34)$$

where $\hbar$ is the Dirac constant and $-e$ is the unit of charge, the charge on the electron accelerated to c. Therefore Eq. (1.1.31) quantizes directly to Eq. (1.1.34), giving the Planck Law,

$$En = \hbar\omega = \hbar c\kappa \ . \qquad (1.1.35)$$

This is the same in *O(3)* and *U(1)* electrodynamics. However, the complete vector A_μ in the internal ((1), (2), (3)) space of *O(3)* is the light-like polar vector,

$$\boldsymbol{A}^{\mu} = \left(0, \boldsymbol{A}^{(1)}\right)\boldsymbol{e}^{(1)} + \left(0, \boldsymbol{A}^{(2)}\right)\boldsymbol{e}^{(2)} + \left(cA^{(0)}, \boldsymbol{A}^{(3)}\right)\boldsymbol{e}^{(3)} , \qquad (1.1.36)$$

and has time-like, longitudinal and transverse components which are each physical. These concepts do not exist in the *U(1)* hypothesis, in which the time-like and longitudinal components are combined to give what is asserted conventionally to be a physical admixture [5].

To summarize, the differences between the *U(1)* and *O(3)* theories are as follows:

1) In *U(1)*, the physical object that we started with was a transverse plane wave with no longitudinal or time-like components. The internal space was a scalar space, and the physical entity was $A^{\mu} = A^{\mu *}$.

2) In *O(3)*, the physical object has become transverse, longitudinal and time-like, and the internal gauge space has become a vector space with *O(3)* rotation group symmetry. This leads directly to the Planck Law through Eq. (1.31), a concept which does not exist in the classical *U(1)* hypothesis. We begin to see advantages in the *O(3)* hypothesis.

1.2.2 Field Tensor from Field Potential

With these definitions, the rules of general gauge field theory can be applied to electrodynamics. The groundwork for this was provided in Vol. 2 of this series, and the fundamental methods are given by Ryder [5]. It is first necessary to define the field tensor in *O(3)* through the field potential. The field tensor is also a vector in the internal *O(3)* gauge space,

$$\boldsymbol{G}^{\mu\nu} = G^{\mu\nu(1)}\boldsymbol{e}^{(1)} + G^{\mu\nu(2)}\boldsymbol{e}^{(2)} + G^{\mu\nu(3)}\boldsymbol{e}^{(3)}, \tag{1.1.37}$$

and the coefficients $G^{\mu\nu(i)}$, $i = 1, 2, 3$, are scalar coefficients of the internal space. They are also antisymmetric tensors in Minkowski space-time.

General gauge field theory for *O(3)* symmetry [1—9] then gives

$$\begin{aligned} \boldsymbol{G}^{\mu\nu(1)*} &= \partial^{\mu}\boldsymbol{A}^{\nu(1)*} - \partial^{\nu}\boldsymbol{A}^{\mu(1)*} - ig\boldsymbol{A}^{\mu(2)} \times \boldsymbol{A}^{\nu(3)}, \\ \boldsymbol{G}^{\mu\nu(2)*} &= \partial^{\mu}\boldsymbol{A}^{\nu(2)*} - \partial^{\nu}\boldsymbol{A}^{\mu(2)*} - ig\boldsymbol{A}^{\mu(3)} \times \boldsymbol{A}^{\nu(1)}, \\ \boldsymbol{G}^{\mu\nu(3)*} &= \partial^{\mu}\boldsymbol{A}^{\nu(3)*} - \partial^{\nu}\boldsymbol{A}^{\mu(3)*} - ig\boldsymbol{A}^{\mu(1)} \times \boldsymbol{A}^{\nu(2)}, \end{aligned} \tag{1.1.38}$$

which is a relation between vectors in the internal space ((1), (2), (3)). The cross product notation is also a vector notation, for example $\boldsymbol{A}^{\mu(2)} \times \boldsymbol{A}^{\nu(3)}$ is a cross product of a vector $\boldsymbol{A}^{\mu(2)}$ with the vector $\boldsymbol{A}^{\nu(3)}$ in the internal space. In forming this cross product, the Greek indices μ and ν are not transmuted, and the complex basis ((1), (2), (3)) is used, so that the terms quadratic in $\boldsymbol{A}$ become natural descriptions of the empirically observable conjugate product. It will be shown that these terms give rise to vacuum polarization and vacuum magnetization in *O(3)* but not in *U(1)* electrodynamics. The definition (1.1.38) is for the free field in regions free of matter and free of charge/current interaction. The scalar coefficient g is a scalar both in the internal gauge space, a symmetry space, and also in Minkowski space-time. In the vacuum it is given by [1—4],

$$g = \frac{\kappa}{A^{(0)}} = \frac{e}{\hbar}, \tag{1.1.39}$$

and is the inverse of the quantum of magnetic flux, $\hbar/e$. Evidently, Eq. (1.1.39) is a fundamental quantum relation for one photon. In field matter interaction g changes in magnitude and is empirically determined through the Verdet constant in the inverse Faraday effect, and the non linear terms in Eq. (1.1.39) (those quadratic in $\boldsymbol{A}$) become negligible under most conditions [18]. The *O(3)* theory then reduces to the same algebraic form as the *U(1)* theory for $G^{\mu\nu(1)} = G^{\mu\nu(2)*}$, *i.e.*, reduces to the homogeneous and inhomogeneous Maxwell equations for the complex conjugate field tensors $G^{\mu\nu(1)}$ and $G^{\mu\nu(2)}$. This is the linear approximation which neglects all non linear optical phenomena such as the inverse Faraday effect. The latter is described through equations for $G^{\mu\nu(3)}$, which is always quadratic in the potential and always non linear. This tensor, $G^{\mu\nu(3)}$, contains only the $\boldsymbol{B}^{(3)}$ field. Self consistently, therefore, the $\boldsymbol{B}^{(3)}$ field is undefined in the linear approximation, which is Maxwell's theory. Note that g is never zero in free space, however, and in this condition the *O(3)* electrodynamics differs fundamentally from its *U(1)* counterpart because in free space the magnitude of the non linear term is the same as those linear in $\boldsymbol{A}$.

The main difference between *O(3)* and *U(1)* in this section are therefore as follows:

1) the field tensor in *U(1)* is well known to be the antisymmetric four-curl:

$$G^{\mu\nu} = \partial^{\mu}A^{\nu} - \partial^{\nu}A^{\mu}, \tag{1.1.40}$$

and there is a scalar internal gauge space, *i.e.*, $G^{\mu\nu}$ is a scalar in this space and an antisymmetric tensor in Minkowski space-time. The field tensor is linear in the field potential, and only transverse components are present in *U(1)*.

2) The field tensor in *O(3)* is a vector in the internal gauge space ((1), (2), (3)) and is non linear in the field potential. It contains the longitudinal and fundamental magnetic flux density component $\boldsymbol{B}^{(3)}$.

1.2.2.1 Details of Equation (1.1.38)

Equation (1.1.38) is a concise description which contains a considerable amount of information about the *O(3)* theory of electromagnetism in free space. This information is obtainable without assuming any form of field equation, and so, details are given in this section of the correct algebraic methods of reduction. Considering for example the equation,

$$\boldsymbol{G}^{\mu\nu(1)*} = \partial^\mu \boldsymbol{A}^{\nu(1)*} - \partial^\nu \boldsymbol{A}^{\mu(1)*} - ig\boldsymbol{A}^{\mu(2)} \times \boldsymbol{A}^{\nu(3)} . \qquad (1.1.41)$$

$$G^{12(1)*} = \partial^1 A^{2(1)*} - \partial^2 A^{1(1)*} - ig\epsilon_{(1)(2)(3)} A^{1(2)} A^{2(3)} , \qquad (1.1.42)$$

This equation consists of components such as, where $\epsilon_{(1)(2)(3)}$ is the Levi Civita symbol, defined by

$$\epsilon_{(1)(2)(3)} := 1 = -\epsilon_{(1)(2)(3)} = \epsilon_{(2)(3)(1)} = \ldots. \qquad (1.1.43)$$

If we now take the vector potential as defined in Section (1.2.1.1), with

$$\partial^\mu := \left(\frac{1}{c}\frac{\partial}{\partial t}, -\nabla \right) , \qquad (1.1.44)$$

then,

$$\begin{aligned} G^{12(1)*} &= \partial^1 A^{2(1)*} - \partial^2 A^{1(1)*} \\ &- ig\left(A^{1(2)} A^{2(3)} - A^{1(3)} A^{2(2)} \right) = 0 . \end{aligned} \qquad (1.1.45)$$

This is a self-consistent result because there is no *Z* component of $G^{\mu\nu(1)*}$, which is defined as transverse. Both the linear and non linear components are zero.

We next consider the element,

$$\begin{aligned} G^{13(1)*} &= \partial^1 A^{3(1)*} - \partial^3 A^{1(1)*} - ig\epsilon_{(1)(2)(3)} A^{1(2)} A^{3(3)} \\ &= \partial^1 A^{3(2)} - \partial^3 A^{1(2)} - ig\left(A^{1(2)} A^{3(3)} - A^{1(3)} A^{3(2)}\right) \\ &= -\left(\partial^3 + igA^{3(3)}\right) A^{1(2)} = -\left(\partial^3 + i\kappa\right) A^{1(2)} , \end{aligned} \tag{1.1.46}$$

where we have used,

$$g = \frac{\kappa}{A^{(0)}} , \qquad A^{3(3)} = A_Z^{(3)} = A^{(0)} . \tag{1.1.47}$$

It can be seen that there are two contributions to the field element $G^{13(2)}$, a magnetic field component:

1) the linear contribution, $-\partial^3 A^{1(2)}$;
2) the non-linear contribution, $-igA^{3(3)} A^{1(2)}$.

In vector notation, Eq. (1.1.46) is a component of,

$$\begin{aligned} 2\boldsymbol{B}^{(1)} &:= \nabla \times \boldsymbol{A}^{(1)} - ig\boldsymbol{A}^{(3)} \times \boldsymbol{A}^{(1)} \\ &= \left(\nabla - ig\boldsymbol{A}^{(3)}\right) \times \boldsymbol{A}^{(1)} \\ &= \nabla \times \boldsymbol{A}^{(1)} - \frac{i}{B^{(0)}} \boldsymbol{B}^{(3)} \times \boldsymbol{B}^{(1)} . \end{aligned} \tag{1.1.48}$$

Furthermore,

$$\partial^3 A^{1(2)} = i\kappa A^{1(2)} , \tag{1.1.49}$$

and so it follows that

$$\boldsymbol{B}^{(1)} = \nabla \times \boldsymbol{A}^{(1)} = -\frac{i}{B^{(0)}} \boldsymbol{B}^{(3)} \times \boldsymbol{B}^{(1)} . \tag{1.1.50}$$

Similarly,

$$\boldsymbol{B}^{(2)} = \nabla \times \boldsymbol{A}^{(2)} = -\frac{i}{B^{(0)}} \boldsymbol{B}^{(2)} \times \boldsymbol{B}^{(3)} . \tag{1.1.51}$$

Therefore the definition of the field tensor in *O(3)* electrodynamics gives the first two components of the B Cyclic Theorem [1—4],

$$\left. \begin{aligned} \boldsymbol{B}^{(3)} \times \boldsymbol{B}^{(1)} &= iB^{(0)}\boldsymbol{B}^{(2)*} \\ \boldsymbol{B}^{(2)} \times \boldsymbol{B}^{(3)} &= iB^{(0)}\boldsymbol{B}^{(1)*} \end{aligned} \right\} , \tag{1.1.52}$$

together with the definition of $\boldsymbol{B}^{(1)}$ and $\boldsymbol{B}^{(2)}$ in terms of the curl of vector potentials $\boldsymbol{A}^{(1)}$ and $\boldsymbol{A}^{(2)}$,

$$\left. \begin{aligned} \boldsymbol{B}^{(1)} &= \nabla \times \boldsymbol{A}^{(1)} , \\ \boldsymbol{B}^{(2)} &= \nabla \times \boldsymbol{A}^{(2)} . \end{aligned} \right\} \tag{1.1.53}$$

It is convenient to write this important result as

$$\boldsymbol{H}(\text{vac.}) = \frac{1}{\mu_0}\boldsymbol{B} - \boldsymbol{M}(\text{vac.}) , \tag{1.1.54}$$

where $\boldsymbol{H}(\text{vac})$ is the vacuum magnetic field strength and μ_0 the permeability in vacuo. The object $\boldsymbol{M}(\text{vac})$ does not exist in *U(1)* electrodynamics and is the *vacuum magnetization*, for example,

$$\boldsymbol{M}^{(1)}(\text{vac}) = -\frac{1}{i\mu_0 B^{(0)}} \boldsymbol{B}^{(3)} \times \boldsymbol{B}^{(1)} . \tag{1.1.55}$$

The objects $\boldsymbol{M}^{(1)}(\text{vac.})$ and $\boldsymbol{M}^{(2)}(\text{vac.})$ depend on the phase-less vacuum magnetic field $\boldsymbol{B}^{(3)}$ and so does not exist as a concept in *U(1)* electrodynamics. The $\boldsymbol{B}^{(3)}$ field itself is defined through

$$\boldsymbol{G}^{\mu\nu(3)*} = \partial^{\mu}\boldsymbol{A}^{\nu(3)*} - \partial^{\nu}\boldsymbol{A}^{\mu(3)*} - ig\boldsymbol{A}^{\mu(1)} \times \boldsymbol{A}^{\nu(2)} , \tag{1.1.56}$$

with (3) aligned for convenience in the Z axis. So by definition, the only non zero components are

$$G^{12(3)*} = -G^{12(3)*} = B_Z^{(3)} . \tag{1.1.57}$$

It follows that

$$B_Z^{(3)} = -ig\left(A^{1(1)}A^{2(2)} - A^{1(2)}A^{2(1)}\right) , \tag{1.1.58}$$

or

$$\boldsymbol{B}^{(3)} = \boldsymbol{B}^{(3)*} = -ig\boldsymbol{A}^{(1)} \times \boldsymbol{A}^{(2)} = -\frac{i}{B^{(0)}}\boldsymbol{B}^{(1)} \times \boldsymbol{B}^{(2)} , \tag{1.1.59}$$

giving the third component of the B Cyclic Theorem, $\boldsymbol{B}^{(1)} \times \boldsymbol{B}^{(2)} = iB^{(0)}\boldsymbol{B}^{(3)*}$, and the vacuum magnetization,

$$\boldsymbol{M}^{(3)*} = -\frac{1}{i\mu_0 B^{(0)}}\boldsymbol{B}^{(1)} \times \boldsymbol{B}^{(2)} . \tag{1.1.60}$$

These results are all absent from *U(1)* electrodynamics, but we know from Section (1.2.1.1) that they are consistent with the plane waves $\boldsymbol{A}^{(1)} = \boldsymbol{A}^{(2)*}$. We shall return to this point later, in the context of the *O(3)* field equations and their linearization. Note that $\boldsymbol{B}^{(3)}$ is always defined through $\boldsymbol{A}^{(1)} \times \boldsymbol{A}^{(2)*}$ and is *not* the curl of $\boldsymbol{A}^{(3)}$. The conjugate product is an observable of magneto-optics and so $\boldsymbol{B}^{(3)}$ is non-zero *empirically* in *O(3)* electrodynamics. In *U(1)* electrodynamics it is rigorously zero, and $\boldsymbol{A}^{(1)} \times \boldsymbol{A}^{(2)}$ in *U(1)* electrodynamics is considered to be an operator with no Z component. This is in clear conflict with vector algebra, in that $\boldsymbol{A}^{(1)} \times \boldsymbol{A}^{(2)}$ is aligned in the (3) axis. For this reason we prefer to develop

O(3) electrodynamics systematically, and reduce it to the Maxwell equations using linearization approximations where applicable.

To summarize what we have found so far, in *O(3)* electrodynamics (hereinafter frequently referred to just as "*O(3)*") the magnetic part of the complete free field is defined as a sum of the curl of a vector potential and a vacuum magnetization. The latter is inherent in the structure of the B Cyclic Theorem [1—4]. In *U(1)* electrodynamics there is no $\boldsymbol{B}^{(3)}$ field by definition (or more accurately, by hypothesis) and in consequence there is no vacuum magnetization in classical *U(1)* electrodynamics. In *O(3)* the $\boldsymbol{B}^{(3)}$ field is always proportional by hypothesis to the conjugate product $\boldsymbol{A}^{(1)} \times \boldsymbol{A}^{(2)}$, which in field matter interaction is an optical observable. The $\boldsymbol{B}^{(3)}$ field is not the curl of a vector potential, and this is a clear departure from the *U(1)* hypothesis of classical electrodynamics. The phase-less $\boldsymbol{B}^{(3)}$ is instead directly proportional in free space to the phase-less $\boldsymbol{A}^{(3)}$ through the scalar relation $B^{(0)} = \kappa A^{(0)}$ [1—4]. These results are obtained self consistently from the definition of the field from the potentials in the *O(3)* gauge theory. We have calculated the field coefficients:

$$
\begin{aligned}
G^{01(2)} &= \left(\partial^0 + igA^{0(3)}\right)A^{1(2)} = -G^{10(2)}, \\
G^{02(2)} &= \left(\partial^0 + igA^{0(3)}\right)A^{2(2)} = -G^{20(2)}, \\
G^{03(2)} &= 0, \\
G^{13(2)} &= -\left(\partial^3 + igA^{3(3)}\right)A^{1(2)} = -G^{31(2)}, \\
G^{23(2)} &= -\left(\partial^3 + igA^{3(3)}\right)A^{2(2)} = -G^{32(2)}, \\
G^{12(2)} &= 0 .
\end{aligned}
\tag{1.1.61}
$$

Similarly ,

$$G^{01(1)} = G^{01(2)*} = \left(\partial^0 - igA^{0(3)}\right)A^{1(1)}, \tag{1.1.62}$$

and so on, and,

$$G^{12(3)*} = -G^{21(3)*} = -ig\left(A^{1(1)}A^{2(2)} - A^{1(2)}A^{2(1)}\right) . \tag{1.1.63}$$

The three field tensors are

$$G^{\mu\nu(1)} = G^{\mu\nu(2)*} = \begin{bmatrix} 0 & -E^{1(1)} & -E^{2(1)} & 0 \\ E^{1(1)} & 0 & 0 & cB^{2(1)} \\ E^{2(1)} & 0 & 0 & -cB^{1(1)} \\ 0 & -cB^{2(1)} & cB^{1(1)} & 0 \end{bmatrix}, \tag{1.1.64}$$

the transverse tensor; and the longitudinal

$$G^{\mu\nu(3)*} = G^{\mu\nu(3)} = \begin{bmatrix} 0 & 0 & 0 & 0 \\ 0 & 0 & -cB^{3(3)} & 0 \\ 0 & cB^{3(3)} & 0 & 0 \\ 0 & 0 & 0 & 0 \end{bmatrix}. \tag{1.1.65}$$

In classical *O(3)* electrodynamics there also exists a vacuum polarization, because the complete electric field strength in the vacuum is given by

$$\left.\begin{aligned} 2\boldsymbol{E}^{(2)} &:= -\frac{\partial \boldsymbol{A}^{(2)}}{\partial t} - igcA^{(0)}\boldsymbol{A}^{(2)} \\ &= -\left(\frac{\partial}{\partial t} + igcA^{(0)}\right)\boldsymbol{A}^{(2)} \\ &= 2\boldsymbol{E}^{(1)*} \end{aligned}\right] . \tag{1.1.66}$$

Using $g = \kappa / A^{(0)}$,

$$\boldsymbol{E}^{(2)} = -\frac{\partial \boldsymbol{A}^{(2)}}{\partial t} = -ic\kappa\boldsymbol{A}^{(2)} = -i\omega\boldsymbol{A}^{(2)}, \tag{1.1.67}$$

and it is convenient to express this result as

$$\frac{1}{\epsilon_0}\boldsymbol{D}^{(2)}(\text{vac.}) = \boldsymbol{E}^{(2)} + \frac{1}{\epsilon_0}\boldsymbol{P}^{(2)}(\text{vac.}), \tag{1.1.68}$$

where $\boldsymbol{D}^{(2)}(\text{vac.})$ is the electric displacement in vacuo and where the vacuum polarization is $\boldsymbol{P}^{(2)}(\text{vac.}) = -i\epsilon_0\omega\boldsymbol{A}^{(2)}$, where ϵ_0 is the vacuum permittivity.

The vacuum polarization is well known to have an analogue in quantum electrodynamics: the photon self energy [5]. This has no classical analogue in *U(1)* electrodynamics, but is clearly defined in *O(3)* electrodynamics. The classical *O(3)* vacuum polarization is transverse and vanishes when $\omega = 0$, so has no meaning in electrostatics. This is consistent with the fact that it is the analogue of photon self energy in quantum electrodynamics. Finally, it is pure transverse, because the hypothetical $\boldsymbol{E}^{(3)}$ field is zero in *O(3)* electrodynamics,

$$\begin{aligned} G^{03(3)*} &= \partial^0 A^{3(3)*} - \partial^3 A^{0(3)*} \\ &- ig\left(A^{0(1)}A^{3(2)} - A^{3(2)}A^{0(1)}\right) = 0, \end{aligned} \tag{1.1.69}$$

and so

$$G^{03(1)} = G^{03(2)} = G^{03(3)} = 0, \tag{1.1.70}$$

in the vacuum. In the presence of field matter interaction this result is no longer true because of the Coulomb field, indicating polarization of matter. Polarization of the vacuum takes place through transverse components only. Again this result is missing from *U(1)* theory.

1.2.3 Field Matter Interaction

In the presence of field matter interaction the *O(3)* field tensor equivalent to that in Eq. (1.1.38) of Section (1.2.2) becomes

$$\frac{1}{\epsilon_0}\boldsymbol{H}^{\mu\nu(i)*} = \boldsymbol{F}^{\mu\nu(i)*} - \frac{1}{\epsilon_0}\boldsymbol{M}^{\mu\nu(i)*}, \tag{1.1.71}$$

where $i = 1, 2, 3$. Here,

$$\left.\begin{array}{l} \boldsymbol{F}^{\mu\nu(i)} := \partial^\mu \boldsymbol{A}^{\nu(i)} - \partial^\nu \boldsymbol{A}^{\mu(i)}, \\ \boldsymbol{M}^{\mu\nu(1)} := i\epsilon_0 g' \boldsymbol{A}^{\mu(2)} \times \boldsymbol{A}^{\nu(3)} \end{array}\right], \tag{1.1.72}$$

in cyclic permutation, with $g' \ll g$ empirically [1—4].

1.2.3.1 Example, the Inverse Faraday Effect

In the inverse Faraday effect we have,

$$\boldsymbol{F}^{\mu\nu(3)*} = \boldsymbol{0}, \tag{1.1.75a}$$

$$\boldsymbol{M}^{\mu\nu(3)*} = i\epsilon_0 g' \boldsymbol{A}^{\mu(1)} \times \boldsymbol{A}^{\nu(2)}. \tag{1.1.75b}$$

Equation (1.1.75a) means that the free space $\boldsymbol{B}^{(3)}$ is zero if we attempt to define it as a conventional *U(1)* four-curl. Equation (1.1.75b) in vector notation is

$$\boldsymbol{M}^{(3)*} = i\epsilon^0 g' \boldsymbol{A}^{(1)} \times \boldsymbol{A}^{(2)}, \tag{1.1.76}$$

which is the empirically observed phase free magnetization of the inverse Faraday effect [1—4]. This is a small effect and so $g^{/} \ll g$ empirically. The factor $g^{/}$ for field matter interaction is much smaller than g in free space. In other words the covariant derivative changes its nature when there is field matter interaction, and loosely speaking, this is "bending of space-time" in the presence of charge, akin to bending of space-time in the presence of mass in general relativity. (Recall that the idea of covariant derivative is borrowed from general relativity.) In general, $g^{/}$ is relativistic, and an example of its development is given in Vol.1 [1]. We see that the inverse Faraday effect plays a central role in *O(3)* electrodynamics, which is able to describe the phenomenon from the basic definition of the field tensor. It follows that

$$\boldsymbol{M}^{(3)*} = -\epsilon_0 \frac{g^{/}}{g} \boldsymbol{B}^{(3)} , \tag{1.1.77}$$

for the inverse Faraday effect, which is therefore a direct observation of $\boldsymbol{B}^{(3)}$. Recall that in *U(1)* electrodynamics,

$$\boldsymbol{A}^{(1)} \times \boldsymbol{A}^{(2)} \left(U(1) \right) = \boldsymbol{0} , \tag{1.1.78}$$

and so *U(1)* gauge field theory as applied to electrodynamics does not describe the inverse Faraday effect. The phenomenological invocation of non zero $\boldsymbol{A} \times \boldsymbol{A}^{*} = \boldsymbol{A}^{(1)} \times \boldsymbol{A}^{(2)}$ [1—4] to describe the inverse Faraday effect in *U(1)* theory therefore leads to a paradox, in that the observable does not exist in *U(1)* gauge field theory by definition. The lowest symmetry in which $\boldsymbol{A}^{(1)} \times \boldsymbol{A}^{(2)*}$ exists is *O(3)* [6,8,9], as argued here. The development of *O(3)* = *SU(2)* electrodynamics leads to several major advantages as described by Barrett [8,9] and elsewhere [1—4].

1.2.3.2 Some Conceptual Similarities to and Differences from Yang Mills Theory in High Energy Physics

There are obvious points of similarity between the *O(3)* theory of electrodynamics and conventional Yang-Mills theory in particle physics. Both theories are based on an *SU(2) = O(3)* Lagrangian and the structure of the field tensor and field equations is fundamentally the same. However, there are some differences also. One of these is that in *O(3)* electrodynamics the presence of the non-linearity preceded by g or g' in the field tensor definition does not mean that the particle concomitant with the gauge field is a charged particle. In *O(3)* electrodynamics, the field does not act as its own source because the nonlinearities in the definition of the field tensor are interpretable as vacuum polarization and magnetization. The g constant in *O(3)* electrodynamics is proportional to the charge e, (the charge on the proton), but it is well known that the electron accelerated to the speed of light takes on the attributes of a classical electromagnetic *field* as argued by Jackson [19]. This does not mean that the field is charged. It is also well known that the vector potential is C negative, and is proportional to e in the vacuum, but again, A^{μ} is not charged.

As argued in Chap. 2, the internal (gauge) space, and space-time in classical electrodynamics are not independent spaces, they form an extended Lie algebra as defined by Aldrovandi [20] and discussed in detail in Chap. 2. In particle theory the internal space is usually ascribed to an isospin which is independent of space-time. Generally, however, the internal gauge space is a symmetry space and the basis ((1), (2), (3)) has *O(3)* symmetry. Finally, the constant g is defined by Eq. (1.1.47) in free space, but in field-matter interaction is much smaller in magnitude, as determined empirically and from phenomenological, or semi-classical, non-linear optical theory [1—4]. In elementary particle theory the parameter g is usually interpreted as a constant. However, the structure of the gauge field theory is the same for elementary particle theory and electrodynamics. If the latter is quantized, g becomes a constant $e/\hbar$ in free space [1—4], and in field-matter interaction becomes a coefficient proportional to $e/\hbar$. Evidently, the elementary charge

e is the same scalar quantity in both *U(1)* and *O(3)*, *i.e.*, the charge on the proton, the negative of the charge on the electron ($-e$).

1.3 Gauge Transformation in *O(3)* Electrodynamics

There is a profound difference between *U(1)* and *O(3)* electrodynamics in respect of gauge transformation, and so it is important to give considerable calculational detail as in this section. In *U(1)* the potential is subsidiary to the field, as argued by Heaviside and contemporaries in the late nineteenth century. It was Heaviside's avowed intention to murder the potential, but in *O(3)* it springs to life again, as we shall find. In *U(1)*, the gauge transformation process is in the last analysis a mathematical convenience, because the gradient of an arbitrary variable is added to the original $\boldsymbol{A}$. This means that gauge transformation of the second kind essentially adds a random quantity to the electromagnetic phase. In non-Abelian gauge field theory applied to classical electrodynamics, the gauge transformation becomes essentially a geometrical process, and there is a well defined topological phase effect [8,9], related to the Aharonov-Bohm effect [8,9]. This is an observed phase effect, and is not random. There are several other features of *O(3)* which do not occur in *U(1)*, and in respect of gauge transformation, the two theories are very different in nature. The main difference is that the potential in *O(3)* and higher symmetry electrodynamic theories is always a physical object, never a mathematical subsidiary variable. In *O(3)*, gauge transformation is controlled by the rules of general gauge field theory as described for example by Ryder [5]. Such ideas form the basis for contemporary gauge field theories such as instanton theory in high energy physics. They are being applied increasingly to low energy physics and to electrodynamics [8,9]. The careful work by Barrett [8,9] in favor of the physical nature of the *classical* electromagnetic potential, and in favor of *SU(2)* = *O(3)* electrodynamics, appears to be irrefutable to the state of the art, based as it is on several different effects of nature. Since $\boldsymbol{A}^{(1)} \times \boldsymbol{A}^{(2)}$ is missing by definition [1—4] from *U(1)* gauge field theory applied to classical electrodynamics ("*U(1)*" for short) the various non-linear magneto-optical effects [18] may be added to the list given by Barrett. If so, it follows that *O(3)* = *SU(2)* symmetry is to be preferred over *U(1)* for a more consistent view of optics, in a theoretical framework which envelops both linear and non-linear phenomena. This is a powerful geometrical argument in favor of *O(3)* because in *U(1)*, the conjugate product $\boldsymbol{A}^{(1)} \times \boldsymbol{A}^{(2)}$ must be an operator with no longitudinal component. This makes no sense in three dimensional space, since by analogy with the longitudinally directed Poynting vector, a cross product of transverse field components; $\boldsymbol{A}^{(1)} \times \boldsymbol{A}^{(2)}$ must also be longitudinally directed for elementary consistency. Similarly, the angular momentum of a classical electromagnetic beam is longitudinally directed, as argued by Jackson [19]. So the *U(1)* appellation in classical electrodynamics can refer at best only to the Lagrangian. In other contexts it is self contradictory as evidenced in the vacuum by the Poynting vector, or angular momentum vectors, both of which are perpendicular to the plane of the O(2) = *U(1)* symmetry group, and both of which are empirical observables in respectively the Lebedev and Beth effects [4]. Similarly for $\boldsymbol{B}^{(3)}$, and *O(3)* is to be preferred to deal with non linear phenomena within gauge field theory. Such phenomena present an Achilles heel of the standard model as discussed here and elsewhere [1—4]. General gauge field theory has been notably successful in elementary particle theory [5], and may be as successful in classical electrodynamics, but with conceptual differences as discussed already. An important difference appears at present to be that the two spaces in *O(3)* are not independent. The *O(3)* hypothesis has the major advantage of being able to incorporate within one structure non-linear and linear phenomena of optics, and also to logically accommodate such quantities as the Poynting vector as just discussed. There is no doubt that this vector is longitudinally directed and outside the plane of definition of O(2) = *U(1)*. It is not consistent to apply O(2) to an energy combination (the Lagrangian) and not to the momentum of the same field, the Poynting vector. In *O(3)* the Lagrangian and momentum have the symmetry of three dimensional space, the internal gauge space.

In order to progress from *U(1)* to *O(3)* the concepts of gauge transformation in general field theory are illustrated in detail in this section to show that the gauge transform process is essentially geometrical. In *U(1)*, the gauge transform is essentially a matter of adding to the magnetic vector potential the gradient of a function which can have any value whatsoever without affecting the original magnetic field. So this is an arbitrary, or random, process in the sense that a random mathematical quantity has no physical meaning unless subjected to thermodynamic averaging. Yet the incorporation of such a quantity is precisely the basis of *U(1)* gauge transformation of the second kind [6,8,9]. In *U(1)*, the electromagnetic phase is random.

1.3.1 The Fundamental Gauge Transform Equations

In the condensed matrix notation used by Ryder [5], the basic equations of gauge transformation in general field theory are as follows

$$G_{\mu\nu} = \frac{i}{g}\left[D_{\mu}, D_{\nu}\right], \tag{1.1.79a}$$

$$G'_{\mu\nu} = S G_{\mu\nu} S^{-1}, \tag{1.1.79b}$$

$$A'_{\mu} = \left(S A_{\mu} - \frac{i}{g}\partial_{\mu} S \right) S^{-1}. \tag{1.1.79c}$$

In this notation, S is a rotation matrix, A_{μ} is a matrix generated from the vector potential, and $G_{\mu\nu}$, the field matrix, is defined as the commutator of covariant derivatives, D_{μ}. Gauge transformation as in Eq. (1.1.79c) is a rotation using curvilinear coordinates, one which changes covariantly. These equations represent physical rotation. If *O(3)*, the rotation group, is used as the background or internal gauge field symmetry of the field theory, the rotation takes place in three dimensions. These ideas have been applied to electromagnetism in previous volumes [1—4], to which the interested reader is referred for more detail. In this section full details of the gauge transform process are given for a rotation about the Z axis.

1.3.2 Background Mathematical Detail

Some background details of the operation of rotation in three dimensional space are given in this section in order to prepare the way for the detailed development of Eqs. (1.1.79a) to (1.1.79c). We consider the Euler angles α, β, and γ and the quaternion coefficients q_0, q_1, q_2 and q_3. Define,

$$\left.\begin{aligned} a &= q_0 + iq_3 = \cos\frac{\beta}{2}\exp\left(\frac{i}{2}(\alpha+\gamma)\right) \\ b &= q_1 - iq_2 = \sin\frac{\beta}{2}\exp\left(\frac{-i}{2}(\alpha-\gamma)\right) \end{aligned}\right\} \tag{1.1.80}$$

with

$$q_0^2 + q_1^2 + q_2^2 + q_3^2 = 1 . \tag{1.1.81}$$

Then the spinor rotation in SU(2) is

$$\begin{bmatrix} u' \\ v' \end{bmatrix} = \begin{bmatrix} a & b \\ -b^* & a^* \end{bmatrix} \begin{bmatrix} u \\ v \end{bmatrix}, \tag{1.1.82}$$

with determinant ± 1 and $ad - bc = 1$. This can be re-expressed directly in terms of quaternions by

$$\begin{bmatrix} u' \\ v' \end{bmatrix} = \begin{bmatrix} q_0 + iq_3 & q_1 - iq_2 \\ -q_1 - iq_2 & q_0 - iq_3 \end{bmatrix} \begin{bmatrix} u \\ v \end{bmatrix} . \tag{1.1.83}$$

In *O(3)*, whose covering group is *SU(2)*, the rotation matrix, is

$$\begin{bmatrix} X' \\ Y' \\ Z' \end{bmatrix} = \begin{bmatrix} e_{1X} & e_{1Y} & e_{1Z} \\ e_{2X} & e_{2Y} & e_{2Z} \\ e_{3X} & e_{3Y} & e_{3Z} \end{bmatrix} \begin{bmatrix} X \\ Y \\ Z \end{bmatrix} , \tag{1.1.84}$$

where $\boldsymbol{e}_1$, $\boldsymbol{e}_2$, and $\boldsymbol{e}_3$ are unit vectors whose components are defined by

$$\left.\begin{aligned} e_{1X} &= q_0^2 + q_1^2 - q_3^2 - q_3^2 = \cos\alpha\,\cos\beta\,\cos\gamma - \sin\alpha\,\sin\beta\,, \\ e_{1Y} &= 2\,(q_1 q_2 + q_0 q_3) = \sin\alpha\,\cos\beta\,\cos\gamma + \cos\alpha\,\sin\gamma\,, \\ e_{1Z} &= 2\,(q_1 q_3 - q_0 q_2) = -\sin\beta\,\cos\gamma\,, \\ e_{2X} &= 2\,(q_1 q_2 - q_0 q_3) = -\cos\alpha\,\cos\beta\,\sin\gamma - \sin\alpha\,\cos\gamma\,, \\ e_{2Y} &= q_0^2 - q_1^2 + q_2^2 - q_3^2 = -\sin\alpha\,\cos\beta\,\sin\gamma + \cos\alpha\,\cos\gamma\,, \\ e_{2Z} &= 2\,(q_2 q_3 + q_0 q_1) = \sin\beta\,\sin\gamma\,, \\ e_{3X} &= 2\,(q_1 q_3 + q_0 q_2) = \cos\alpha\,\sin\beta\,, \\ e_{3Y} &= 2\,(q_2 q_3 - q_0 q_1) = -\sin\alpha\,\sin\beta\,, \\ e_{3Z} &= q_0^2 - q_1^2 - q_2^2 + q_3^2 = \cos\beta\,. \end{aligned}\right\} \tag{1.1.85}$$

Therefore rotation in three dimensions can be represented equivalently in terms of vectors, spinors, quaternions, and Euler angles. Rotation about the Z axis is represented by

$$\cos\beta = \cos\gamma = 1\,, \qquad \sin\beta = \sin\gamma = 0\,, \tag{1.1.86}$$

and so

$$\left.\begin{array}{lll} e_{1X} = \cos\alpha\,, & e_{1Y} = \sin\alpha\,, & e_{1Z} = 0\,, \\ e_{2X} = -\sin\alpha\,, & e_{2Y} = \cos\alpha\,, & e_{2Z} = 0\,, \\ e_{3X} = 0\,, & e_{3Y} = 0\,, & e_{3Z} = 1\,. \end{array}\right\} \tag{1.1.87}$$

A possible description of rotation about the Z axis is $\beta = 0$, $\gamma = 0$, *i.e.*,

$$q_0 = \cos\frac{\alpha}{2}\,, \qquad q_3 = \sin\frac{\alpha}{2}\,, \qquad q_1 = 0\,, \qquad q_2 = 0\,. \tag{1.1.88}$$

The rotation matrices are therefore

$$\begin{bmatrix} \cos\alpha & \sin\alpha & 0 \\ -\sin\alpha & \cos\alpha & 0 \\ 0 & 0 & 1 \end{bmatrix} \Leftrightarrow \begin{bmatrix} e^{i\alpha/2} & 0 \\ 0 & e^{-i\alpha/2} \end{bmatrix}, \tag{1.1.89}$$

or, in terms of quaternion components or coefficients

$$\begin{bmatrix} q_0^2 - q_3^2 & 2q_0q_3 & 0 \\ -2q_0q_3 & q_0^2 - q_3^2 & 0 \\ 0 & 0 & 1 \end{bmatrix} \Leftrightarrow \begin{bmatrix} q_0 + iq_3 & 0 \\ 0 & q_0 - iq_3 \end{bmatrix}. \tag{1.1.90}$$

The self consistency of this process can be checked through the fact that it gives the well known half angle formulae,

$$\begin{aligned} \cos\alpha &= q_0^2 - q_3^2 = \cos^2\frac{\alpha}{2} - \sin^2\frac{\alpha}{2}\,, \\ \sin\alpha &= 2q_0q_3 = 2\cos\frac{\alpha}{2}\sin\frac{\alpha}{2}\,. \end{aligned} \tag{1.1.91}$$

Note that the *O(3)* rotation matrix is set up in terms of α and the *SU(2)* in terms of $\alpha/2$. The *O(3)* rotation matrix is real, the *SU(2)* rotation matrix is complex. The same quaternion coefficients appear in *O(3)* and *SU(2)*.

1.3.3 Infinitesimal Rotation Generator in *SU(2)*

Our first example of the development of Eqs. (1.1.79a) to (1.1.79c) uses infinitesimal rotation generators in *SU(2)*. Let

$$R_\alpha(Z) := \begin{bmatrix} e^{i\alpha/2} & 0 \\ 0 & e^{-i\alpha/2} \end{bmatrix}, \tag{1.1.92}$$

be an *SU(2)* rotation matrix. Its infinitesimal rotation generator is then defined to be

$$\tau_Z := \frac{1}{i}\frac{\partial R_Z}{\partial \alpha}(\alpha)\bigg|_{\alpha=0} = \frac{1}{2}\begin{bmatrix} 1 & 0 \\ 0 & -1 \end{bmatrix} := \frac{\sigma_Z}{2}, \tag{1.1.93}$$

where σ_Z is the third Pauli matrix.

Now apply the Taylor series to the matrix exponential to obtain

$$e^{i\sigma_Z\alpha/2} = 1 + i\sigma_Z\frac{\alpha}{2} - \frac{\sigma_Z^2}{2!}\frac{\alpha^2}{4} - i\frac{\sigma_Z^3}{3!}\frac{\alpha^3}{8} + \ldots$$

$$= \begin{bmatrix} 1 & 0 \\ 0 & 1 \end{bmatrix} + i\begin{bmatrix} 1 & 0 \\ 0 & -1 \end{bmatrix}\frac{\alpha}{2} - \frac{1}{2!}\begin{bmatrix} 1 & 0 \\ 0 & 1 \end{bmatrix}\frac{\alpha^2}{4} + \ldots. \tag{1.1.94}$$

$$= \begin{bmatrix} e^{i\alpha/2} & 0 \\ 0 & e^{-i\alpha/2} \end{bmatrix},$$

and therefore

$$R_\alpha(Z) = e^{i\sigma_Z\alpha/2} = q_0 + iq_3q_Z. \tag{1.1.95}$$

In the small angle limit,

$$q_0 \longrightarrow 1, \qquad q_3 \longrightarrow \frac{\alpha}{2}, \tag{1.1.96}$$

and

$$e^{i\sigma_Z\alpha/2} \underset{\alpha \to 0}{\longrightarrow} 1 + i\frac{\alpha}{2}\sigma_Z, \tag{1.1.97}$$

which are self consistently the first two terms of a Taylor series.

1.3.3.1 Field Rotation in *SU(2)*

The rotation of a field ψ is now definable by [5],

$$\psi' = e^{i\sigma_Z\alpha/2}\psi = \left(q_0 + iq_3\sigma_2\right)\psi, \tag{1.1.98}$$

and with these components in hand the gauge transformation process in *SU(2)* is based on the idea that the Euler angle α is a function of x^μ, the space-time four-vector. This is a gauge transformation of the second kind, which is underpinned by special relativity [1—5]. The quaternion coefficients become functions of x^μ, and derivatives are replaced by covariant derivatives in *SU(2)* [5]. Under gauge transformation of the second kind, the potential four-vector becomes

$$A'_\mu = SA_\mu S^{-1} - \frac{i}{g}\partial_\mu SS^{-1}, \tag{1.1.99}$$

in which appears an inhomogeneous, purely topological, term, the second term on the right hand side. In our example, this equation is developed as

$$A_\mu := A_\mu^a \frac{\sigma^a}{2} = A_\mu^Z \frac{\sigma^Z}{2}, \qquad S := e^{i\sigma_Z \alpha/2}. \tag{1.1.100}$$

In analogy with Eq. (1.1.1) of Sec. 1.2, the object A_μ is a matrix in an internal gauge space indicated by the superscript a. In this notation, summation is implied over all repeated indices. Greek indices are covariant-contravariant Minkowski space indices. Latin ones denote the internal gauge space. The placement of the Latin indices as subscript or superscript is not significant, because they are not contravariant-covariant indices. For the rotation about the Z axis that we are considering here, $a = Z$. The symbol S is a rotation matrix in *SU(2)* in exponential form. Therefore the symbol A_μ is interpreted as the matrix,

$$A_\mu := \begin{bmatrix} \frac{A_\mu^Z}{2} & 0 \\ 0 & -\frac{A_\mu^Z}{2} \end{bmatrix}, \tag{1.1.101}$$

for this example of Z axis rotation in *SU(2)* of the field ψ. The *SU(2)* gauge transformation of A_μ is given by Eq. (1.1.99), with its characteristic inhomogeneous or topological term. In a *U(1)* symmetry theory this term is the well known gradient of an arbitrary function first introduced in the late nineteenth century. In *SU(2)* however, it is not arbitrary, and is determined by S, *i.e.*, by a particular Euler angle α, or quaternion component.

1.3.3.2 The Inhomogeneous or Topological Term in *SU(2)*

We use Eq. (1.1.99) with

$$S = e^{i\alpha\sigma_{Z/2}}, \qquad S^{-1} = e^{-i\alpha\sigma_Z/2}, \tag{1.1.102}$$

and

$$\partial_\mu S = \left(i\frac{\sigma_Z}{2}\partial_\mu\alpha \right) S. \tag{1.1.103}$$

Therefore gauge transformation results in

$$A_\mu^{/} = A_\mu - \frac{i^2}{g}\frac{\sigma_Z}{2}\partial_\mu\alpha, \tag{1.1.104}$$

or in matrix form,

$$\begin{bmatrix} \frac{A_\mu^{Z'}}{2} & 0 \\ 0 & -\frac{A_\mu^{Z'}}{2} \end{bmatrix} = \begin{bmatrix} \frac{A_\mu^{Z}}{2} & 0 \\ 0 & -\frac{A_\mu^{Z}}{2} \end{bmatrix} + \frac{1}{2g}\begin{bmatrix} \partial_\mu\alpha & 0 \\ 0 & -\partial_\mu\alpha \end{bmatrix}, \tag{1.1.105}$$

where,

$$\alpha = \cos^{-1}\left(q_0^2 - q_3^2\right) = \sin^{-1}\left(2q_0q_3\right) . \tag{1.1.106}$$

Therefore ,

$$A_\mu^{Z'} = A_\mu^{Z} + \frac{1}{g}\partial_\mu\alpha . \tag{1.1.107}$$

This is clearly a geometrical result, rotation of the field ψ about the *Z* axis has this effect on the *Z* component of A_μ in the *SU(2)* internal gauge space. We are dealing with curvilinear coordinates because in a flat space-time, $\partial_\mu\alpha = 0$ because α is not a function of x^μ. Terms such as $(1/g)\partial_\mu\alpha$ are the physical bases of effects such as that of Aharonov and Bohm. The latter are usually given in terms of *U(1)* electrodynamics, in which α is effectively an arbitrary function. In *SU(2)*, α is clearly the Euler angle, and a finite rotation must always take place through a finite Euler angle.

In the small angle limit,

$$\frac{\alpha}{2} \sim \sin\frac{\alpha}{2} = q_3 , \tag{1.1.108}$$

and so,

$$A_\mu^{Z'} \underset{\alpha \to 0}{\longrightarrow} A_\mu^{Z} + \frac{1}{g}\partial_\mu q_3 . \tag{1.1.109}$$

Note that gauge transformation in an *SU(2)* symmetry field theory is a geometrical process. If $\partial\alpha/\partial x^\mu = 0$, A_Z' goes to A_Z, there is no topological term and no Aharonov-Bohm effect. The object A_μ is a physical four-potential in the *classical* field theory. It is not a mathematical subsidiary variable as in a *U(1)* gauge field theory of classical electrodynamics. There is therefore a profound difference between *O(3)* and *U(1)* electrodynamics.

1.3.3.3 Self Consistency of Equation (1.1.107)

There are various ways of self checking Eq. (1.1.107), for example, for small angle rotation in the *O(3)* group, homomorphic [5,8,9] to *SU(2)*, we should obtain the same result. It is convenient to develop the concise description given by Ryder on his p. 119 [5], and to consider the small angle rotation of a field ϕ with components described by ϕ_1, ϕ_2, and ϕ_3, in general, a matter field. In the *O(3)* internal space,

$$\begin{bmatrix} \phi_1' \\ \phi_2' \\ \phi_3' \end{bmatrix} = \begin{bmatrix} 1 & \Lambda_3 & 0 \\ -\Lambda_3 & 1 & 0 \\ 0 & 0 & 1 \end{bmatrix} \begin{bmatrix} \phi_1 \\ \phi_2 \\ \phi_3 \end{bmatrix}, \tag{1.1.110}$$

for a rotation about the small angle Λ_3. This process is

$$\left.\begin{aligned} \phi_1' &= \phi_1 + \Lambda_3\phi_2 , \\ \phi_2' &= \phi_2 - \Lambda_3\phi_1 , \\ \phi_3' &= \phi_3 , \end{aligned}\right\} \tag{1.1.111}$$

and is a component of the small angle rotation given by $-\Lambda \times \phi$. When $\Lambda = \Lambda_3 \boldsymbol{k}$ we obtain, self consistently,

$$-\Lambda \times \phi = - \begin{vmatrix} \boldsymbol{i} & \boldsymbol{j} & \boldsymbol{k} \\ 0 & 0 & \Lambda_3 \\ \phi_1 & \phi_2 & \phi_3 \end{vmatrix} = \Lambda_3\phi_2 \boldsymbol{i} - \Lambda_3\phi_1 \boldsymbol{j} . \tag{1.1.112}$$

In *O(3)* vector notation,

$$\phi' = e^{i\boldsymbol{J}\cdot\boldsymbol{\Lambda}}\,\boldsymbol{\phi} \Leftrightarrow \boldsymbol{\phi}' = \boldsymbol{\phi} - \boldsymbol{\Lambda}\times\boldsymbol{\phi}\,, \tag{1.1.113}$$

in the small angle limit.

Now apply the formula for gauge transformation,

$$A'_\mu = \left(SA_\mu - \frac{i}{g}\partial_\mu S \right) S^{-1}\,, \tag{1.1.114}$$

with

$$SA_\mu = \exp\left(i\boldsymbol{J}\cdot\boldsymbol{\Lambda}\right)\boldsymbol{A}_\mu \sim \boldsymbol{A}_\mu - \boldsymbol{\Lambda}\times\boldsymbol{A}_\mu\,, \tag{1.1.115a}$$

where $\boldsymbol{A}_\mu$ is a vector in the internal *O(3)* group space, with

$$\partial_\mu S = \left(i\partial_\mu\Lambda\right)S\,, \tag{1.1.115b}$$

to obtain

$$\boldsymbol{A}'_\mu = \left(\boldsymbol{A}_\mu - \boldsymbol{\Lambda}\times\boldsymbol{A}_\mu - \frac{i^2}{g}\partial_\mu\Lambda S \right) S^{-1}\,, \tag{1.1.116}$$

where

$$\left.\begin{aligned} S &= e^{i\boldsymbol{J}\cdot\boldsymbol{\Lambda}} = 1 + i\boldsymbol{J}\cdot\boldsymbol{\Lambda} + \ldots. \\ S^{-1} &= e^{-i\boldsymbol{J}\cdot\boldsymbol{\Lambda}} = 1 - i\boldsymbol{J}\cdot\boldsymbol{\Lambda} + \ldots. \end{aligned}\right\} \tag{1.1.117}$$

so

$$\boldsymbol{A}'_\mu \sim \boldsymbol{A}_\mu - \boldsymbol{\Lambda}\times\boldsymbol{A}_\mu + \frac{1}{g}\partial_\mu\boldsymbol{\Lambda} + \ \ldots\ , \tag{1.1.118}$$

which is the Yang-Mills approximation given by Ryder. The small angle gauge transformation in the *O(3)* gauge group's internal space is a geometrical process, not a random process as in the *U(1)* gauge group. Later an example of this different role played by the potential is considered, the gauge transformation of the conjugate product $\boldsymbol{A}^{(1)} \times \boldsymbol{A}^{(2)}$. In *O(3)* this object is physical, in *U(1)* it is unphysical. However, it is an observable of magneto-optics, and so empirical data prefer the *O(3)* hypothesis. In *O(3)*, rotation about the *Z* axis, a gauge transformation, leaves $\boldsymbol{A}^{(1)} \times \boldsymbol{A}^{(2)}$ unchanged; in *U(1)*, it becomes random, because $\boldsymbol{A}^{(1)} = \boldsymbol{A}^{(2)*}$ becomes random..

Returning to the development in this section, then for a *Z* axis rotation,

$$\Lambda_1 = \Lambda_2 = 0\,, \tag{1.1.119}$$

and

$$-\boldsymbol{\Lambda} \times \boldsymbol{A}_\mu = \Lambda_3 A_{\mu 2}\,\boldsymbol{i} - \Lambda_3 A_{\mu 1}\,\boldsymbol{j}\,, \tag{1.1.120}$$

where $\boldsymbol{i}$, $\boldsymbol{j}$ and $\boldsymbol{k}$ are Cartesian unit vectors in the internal space. So,

$$\left.\begin{aligned} A'_{\mu 1} &= A_{\mu 1} + \Lambda_3 A_{\mu 2}\,, \\ A'_{\mu 2} &= A_{\mu 2} - \Lambda_3 A_{\mu 1}\,, \\ A'_{\mu 3} &= A_{\mu 3} + \frac{1}{g}\partial_\mu \Lambda_3\,. \end{aligned}\right\} \tag{1.1.121}$$

The third of these equations is Eq (1.1.107) in the small angle limit, *QED*.

1.3.4 Gauge Transformation in an *O(3)* Gauge Field Theory

Considering a Z axis rotation in an internal *O(3)* space of a gauge field theory governed by Eqs (1.1.79a) to (1.1.79c) we obtain,

$$S = e^{iJ_Z\alpha}, \qquad S^{-1} = e^{-iJ_Z\alpha}, \tag{1.1.122}$$

where J_Z is the infinitesimal rotation generator defined in Ref 5. Thus,

$$S = 1 + iJ_Z\alpha - J_Z^2\frac{\alpha^2}{2!} - iJ_Z^3\frac{\alpha^3}{3!} + \ldots = \begin{bmatrix} \cos\alpha & \sin\alpha & 0 \\ -\sin\alpha & \cos\alpha & 0 \\ 0 & 0 & 1 \end{bmatrix}, \tag{1.1.123}$$

which is self-consistently the rotation matrix for a rotation about the Z axis in an *O(3)* symmetry gauge field theory.

The inverse of S is formed by $\alpha \rightarrow -\alpha$,

$$S^{-1} = \begin{bmatrix} \cos\alpha & -\sin\alpha & 0 \\ \sin\alpha & \cos\alpha & 0 \\ 0 & 0 & 1 \end{bmatrix} = e^{-iJ_z\alpha}, \tag{1.1.124}$$

and it is easily checked that SS^{-1} is the unit 3 x 3 matrix as required.

The existence of the term $\partial_\mu S$ depends on α being a function of x^μ, since α is the only independent variable in S. So,

$$\partial_\mu S = \partial_\mu \begin{bmatrix} \cos\alpha & \sin\alpha & 0 \\ -\sin\alpha & \cos\alpha & 0 \\ 0 & 0 & 1 \end{bmatrix}, \tag{1.1.125}$$

Now use the calculus,

$$\frac{dy}{dx} = \frac{df}{dx}\frac{dy}{df}, \tag{1.1.126}$$

so if $y = \cos\left(f(x)\right)$ for example, then,

$$\frac{dy}{dx} = -f'(x)\sin\left(f(x)\right). \tag{1.1.127}$$

We obtain

$$\left.\begin{aligned}\partial_\mu\left(\cos\alpha(x^\mu)\right) &= -\partial_\mu\alpha\sin\alpha\,,\\ \partial_\mu\left(\sin\alpha(x^\mu)\right) &= \partial_\mu\alpha\cos\alpha\,,\end{aligned}\right] \tag{1.1.128}$$

and

$$\partial_\mu S = \partial_\mu\alpha\begin{bmatrix}-\sin\alpha & \cos\alpha & 0\\ -\cos\alpha & -\sin\alpha & 0\\ 0 & 0 & 0\end{bmatrix}. \tag{1.1.129}$$

The existence of $\partial_\mu S$ depends directly on that of $\partial_\mu\alpha$ and on the postulate that α is a function of x^μ; a postulate that springs directly from special relativity via type two gauge transform theory [5], or gauge transformation of the second kind.

1.3.4.1 Definition of A_μ

In *O(3)* symmetry gauge field theory the object A_μ is expressed as a matrix,

$$A_\mu = J^a A_\mu^a, \tag{1.1.130}$$

where J^a are the three infinitesimal rotation generator matrices of *O(3)* [1—9] and where the double indexed A_μ^a are scalar coefficients of the internal space, a vector space. For *Z* axis rotation,

$$A_\mu = J^Z A_\mu^Z. \tag{1.1.131}$$

In this notation, the placing of *Z* as an upper or lower index has no algebraic significance, as discussed already, whereas μ is covariant-contravariant. Thus, for *Z* axis rotation,

$$A_\mu = \begin{bmatrix} 0 & -i & 0 \\ i & 0 & 0 \\ 0 & 0 & 0 \end{bmatrix} A_\mu^Z, \tag{1.1.132}$$

The inhomogeneous term in Eq (1.1.114) is also directly dependent on the existence of $\partial_\mu \alpha$,

$$\partial_\mu S S^{-1} = \partial_\mu \alpha \begin{bmatrix} -\sin\alpha & \cos\alpha & 0 \\ -\cos\alpha & -\sin\alpha & 0 \\ 0 & 0 & 0 \end{bmatrix} \begin{bmatrix} \cos\alpha & -\sin\alpha & 0 \\ \sin\alpha & \cos\alpha & 0 \\ 0 & 0 & 1 \end{bmatrix}$$

$$= \partial_\mu \alpha \begin{bmatrix} 0 & 1 & 0 \\ -1 & 0 & 0 \\ 0 & 0 & 0 \end{bmatrix} . \tag{1.1.133}$$

So,

$$-\frac{i}{g}\partial_\mu S S^{-1} = \frac{J_Z}{g}\partial_\mu \alpha . \tag{1.1.134}$$

Note that this is the topological term responsible for the Aharonov-Bohm effect and so forth [1—9]. The scalar g is a dimensionality coefficient introduced as such in the definition of the covariant derivative [5]. The operator J_Z is the infinitesimal rotation generator of *O(3)* about *Z*. The existence of this term in the gauge transform of A_μ is the direct result of special relativity, of gauge transformation of the second kind. In a *U(1)* gauge field theory the equivalent of α is arbitrary, and has no geometrical meaning as we have argued already. In the *O(3)* = *SU(2)* version it is an Euler angle which is a function of x^μ for a given rotation, $\alpha(x^\mu)$ is clearly finite and well defined, being a physical Euler angle in curvilinear coordinates necessitated by special relativity.

The above calculation can be checked for self consistency using the operator formalism. If,

$$S = \exp\left(iJ_Z\alpha\right), \qquad \text{then} \qquad \partial_\mu S = iJ_Z \partial_\mu \alpha S, \tag{1.1.135}$$

and, QED,

$$\partial_\mu S S^{-1} = i J_Z \partial_\mu \alpha \,. \tag{1.1.136}$$

1.3.4.2 The Term $SA_\mu S^{-1}$

This is also a matrix given by,

$$SA_\mu S^{-1} = -iA_\mu^Z$$

$$\times \begin{bmatrix} \cos\alpha & \sin\alpha & 0 \\ -\sin\alpha & \cos\alpha & 0 \\ 0 & 0 & 1 \end{bmatrix} \begin{bmatrix} 0 & 1 & 0 \\ -1 & 0 & 0 \\ 0 & 0 & 0 \end{bmatrix} \begin{bmatrix} \cos\alpha & -\sin\alpha & 0 \\ \sin\alpha & \cos\alpha & 0 \\ 0 & 0 & 1 \end{bmatrix} \tag{1.1.137}$$

$$= A_\mu^Z J_Z = A_\mu \,.$$

The overall result of the gauge transformation is therefore,

$$A_\mu^Z J_Z \rightarrow \left(A_\mu^Z + \frac{1}{g} \partial_\mu \alpha \right) J_Z , \tag{1.1.138}$$

i.e.,

$$A_\mu^Z \rightarrow A_\mu^Z + \frac{1}{g} \partial_\mu \alpha \,. \tag{1.1.139}$$

Self consistently, this is Eq. (1.1.107) of Sec. 1.3.3.1. The *O(3)* and *SU(2)* symmetry theories give the same result for the scalar A_μ^Z of the internal gauge space. If the space is such that α has no dependence on x^μ, the A_μ^Z is unchanged by rotation about *Z*. Self-consistently, this is Euclidean space, in which rotation about Z does not change the direction or magnitude of a vector component aligned in Z.

1.3.5 Transformation of the Field Tensor

The rule for transformation of the field tensor in general gauge field theory is,

$$G^{\prime}_{\mu\nu} = S G_{\mu\nu} S^{-1} . \tag{1.1.140}$$

The inhomogeneous term does not appear and the transformation takes place covariantly rather than invariantly as in *U(1)* [5]. The $\boldsymbol{B}^{(3)}$ field transforms as follows, for a Z axis rotation and in matrix algebra,

$$\begin{bmatrix} 0 & -B_Z & 0 \\ -B_Z & 0 & 0 \\ 0 & 0 & 0 \end{bmatrix}$$

$$\rightarrow \begin{bmatrix} \cos\alpha & \sin\alpha & 0 \\ -\sin\alpha & \cos\alpha & 0 \\ 0 & 0 & 1 \end{bmatrix} \begin{bmatrix} 0 & -B_Z & 0 \\ B_Z & 0 & 0 \\ 0 & 0 & 0 \end{bmatrix} \begin{bmatrix} \cos\alpha & -\sin\alpha & 0 \\ \sin\alpha & \cos\alpha & 0 \\ 0 & 0 & 1 \end{bmatrix} \tag{1.1.141}$$

$$= \begin{bmatrix} 0 & -B_Z & 0 \\ B_Z & 0 & 0 \\ 0 & 0 & 0 \end{bmatrix},$$

i.e.,

$$B_Z \rightarrow B_Z . \tag{1.1.142}$$

The $\boldsymbol{B}^{(3)}$ field is therefore self-consistently invariant under rotation about the Z axis, and the *O(3)* gauge transform is a rotation which produces,

$$\left.\begin{aligned} A_Z &\rightarrow A_Z + \frac{1}{g}\partial_Z \alpha\,, \\ B_Z &\rightarrow B_Z\,. \end{aligned}\right\} \tag{1.1.143}$$

In *U(1)* these concepts do not arise because both A_Z and B_Z are zero, and the idea of gauge transformation being a rotation through a physical Euler angle does not exist.

1.3.6 *O(3)* Gauge Transformation of the Optical Conjugate Product $\boldsymbol{A}^{(1)} \times \boldsymbol{A}^{(2)}$

The optical conjugate product is a well accepted physical observable of the semi classical, phenomenological, theory of non-linear optics [1—4]. As argued in several ways [1—4] already this observable is identically zero by definition in *U(1)*. In *O(3)* it is identically non-zero by definition and proportional to $\boldsymbol{B}^{(3)}$ by definition. To check the consistency of the result (1.1.142) of the preceding section this section is devoted to the details of gauge transformation of $\boldsymbol{A}^{(1)} \times \boldsymbol{A}^{(2)}$ in *O(3)*. Since $\boldsymbol{B}^{(3)}$ is invariant under *O(3)* gauge transformation defined as a rotation about Z, so should be $\boldsymbol{A}^{(1)} \times \boldsymbol{A}^{(2)}$. In order for this to be so, we shall see that the gauge transformation in *O(3)* must generate an electromagnetic phase shift defined in terms of the physical angle of rotation. This result is akin to the topological phase [8,9] and the Aharonov-Bohm effect [8,9] as discussed lucidly by Barrett. It means that there exists an *optical* Aharonov-Bohm effect which is measurable in principle by this phase shift. In *U(1)*, as argued already, the electromagnetic phase is random because of the random nature of gauge transformation of the second kind in *U(1)* [1—4]. To see this result in *O(3)*, the gauge transformation rules must be applied carefully to $\boldsymbol{A}^{(1)}$ and to $\boldsymbol{A}^{(2)}$ as follows,

$$\left.\begin{aligned} A^{(1)} &\rightarrow SA^{(1)}S^{-1} - \frac{i}{g}\partial_\mu SS^{-1}, \\ A^{(2)} &\rightarrow SA^{(2)}S^{-1} + \frac{i}{g}\left(\partial_\mu SS^{-1}\right)^{*}. \end{aligned}\right] \tag{1.1.144}$$

In vector notation, the $\boldsymbol{A}^{(1)}$ and $\boldsymbol{A}^{(2)}$ components are complex conjugates such as,

$$\begin{aligned} \boldsymbol{A}^{(1)} &= \frac{A^{(0)}}{\sqrt{2}}(\boldsymbol{i} - i\boldsymbol{j})\, e^{i\phi}, & \boldsymbol{A}^{(2)} &= \frac{A^{(0)}}{\sqrt{2}}(\boldsymbol{i} + i\boldsymbol{j})\, e^{-i\phi}, \\ &:= A_X^{(1)}\boldsymbol{i} + A_Y^{(1)}\boldsymbol{j}, & &:= A_X^{(2)}\boldsymbol{i} + A_Y^{(2)}\boldsymbol{j}. \end{aligned} \tag{1.1.145}$$

Therefore in matrix form,

$$\left.\begin{aligned} A^{(1)} &= A_X^{(1)}\begin{bmatrix} 0 & 0 & 0 \\ 0 & 0 & -i \\ 0 & i & 0 \end{bmatrix} + A_Y^{(1)}\begin{bmatrix} 0 & 0 & i \\ 0 & 0 & 0 \\ -i & 0 & 0 \end{bmatrix}, \\ A^{(2)} &= A_X^{(2)}\begin{bmatrix} 0 & 0 & 0 \\ 0 & 0 & -i \\ 0 & i & 0 \end{bmatrix} + A_Y^{(2)}\begin{bmatrix} 0 & 0 & i \\ 0 & 0 & 0 \\ -i & 0 & 0 \end{bmatrix}. \end{aligned}\right\} \tag{1.1.146}$$

Rotation of these terms about the Z axis produces results such as the following,

$$SA_X^{(1)}S^{-1}$$

$$= \begin{bmatrix} \cos\alpha & \sin\alpha & 0 \\ -\sin\alpha & \cos\alpha & 0 \\ 0 & 0 & 1 \end{bmatrix} \begin{bmatrix} 0 & 0 & 0 \\ 0 & 0 & -i \\ 0 & i & 0 \end{bmatrix} \begin{bmatrix} \cos\alpha & -\sin\alpha & 0 \\ \sin\alpha & \cos\alpha & 0 \\ 0 & 0 & 1 \end{bmatrix} A_X^{(1)} \qquad (1.1.147)$$

$$= \begin{bmatrix} 0 & 0 & -i\sin\alpha \\ 0 & 0 & -i\cos\alpha \\ i\sin\alpha & i\cos\alpha & 0 \end{bmatrix} A_X^{(1)} .$$

Therefore the vector $\boldsymbol{A}^{(1)}$ is changed by an *O(3)* gauge transformation defined as a rotation about the *Z* axis. This is self consistent because $\boldsymbol{A}^{(1)}$ has *X* and *Y* components only.

Similarly,

$$SA_Y^{(1)}S^{-1} = A_Y^{(1)} \begin{bmatrix} 0 & 0 & i\cos\alpha \\ 0 & 0 & -i\sin\alpha \\ -i\cos\alpha & i\sin\alpha & 0 \end{bmatrix} . \qquad (1.1.148)$$

It can now be checked that the commutator $[\, SA_X^{(1)}S^{-1} ,\, SA_Y^{(1)}S^{-1} \,]$, is a *Z* axis rotation as required,

$$\left[SA_X^{(1)}S^{-1}, SA_Y^{(1)}S^{-1} \right] = \left(SA_X^{(1)}S^{-1}SA_Y^{(1)}S^{-1} - SA_Y^{(1)}S^{-1}SA_X^{(1)}S^{-1} \right)$$

$$= A_X^{(1)}A_Y^{(1)} \begin{bmatrix} 0 & 0 & -i\sin\alpha \\ 0 & 0 & -i\cos\alpha \\ i\sin\alpha & i\cos\alpha & 0 \end{bmatrix} \begin{bmatrix} 0 & 0 & i\cos\alpha \\ 0 & 0 & -i\sin\alpha \\ -i\cos\alpha & i\sin\alpha & 0 \end{bmatrix}$$

$$-A_X^{(1)}A_Y^{(1)} \begin{bmatrix} 0 & 0 & i\cos\alpha \\ 0 & 0 & -i\sin\alpha \\ -i\cos\alpha & i\sin\alpha & 0 \end{bmatrix} \begin{bmatrix} 0 & 0 & -i\sin\alpha \\ 0 & 0 & -i\cos\alpha \\ i\sin\alpha & i\cos\alpha & 0 \end{bmatrix} \tag{1.1.149}$$

$$= A_X^{(1)}A_Y^{(1)} \begin{bmatrix} 0 & 1 & 0 \\ -1 & 0 & 0 \\ 0 & 0 & 0 \end{bmatrix} = iA_X^{(1)}A_Y^{(1)}J_Z .$$

The overall result is that a rotation about the Z axis changes the X and Y components of the potentials $\boldsymbol{A}^{(1)}$ and $\boldsymbol{A}^{(2)}$, but leaves $\boldsymbol{B}^{(3)}$ unchanged. However, the polar longitudinal component A_Z, (which has no existence in *U(1)*), is changed by the same gauge transform process to $A_Z + \partial_Z\alpha$. Therefore, $\boldsymbol{A}^{(1)} \times \boldsymbol{A}^{(2)}$ is self consistently proportional to $\boldsymbol{B}^{(3)}$ in *O(3)*. In *U(1)*, as we have seen, $\boldsymbol{B}^{(3)}$ is zero and $\boldsymbol{A}^{(1)} \times \boldsymbol{A}^{(2)}$ is randomized by the *U(1)* gauge transformation of the second kind because random quantities are added to $\boldsymbol{A}^{(1)}$ and $\boldsymbol{A}^{(2)}$ (gradients of arbitrary scalars). It seems clear that *O(3)* is the more consistent theory on these arguments alone, because $\boldsymbol{A}^{(1)} \times \boldsymbol{A}^{(2)}$ is an optical *observable*. In *U(1)*, the potential is never an observable according to the Heaviside interpretation, it is strictly a mathematical subsidiary. The latter conclusion has been shown to be false by Barrett [8,9] using half a dozen phenomena of nature. The Heaviside view was criticized by Ritz as early as 1908 [13] on the grounds that the classical potential denoted delayed action at a distance as advocated by Schwarzschild in 1902 [12], and so must be physical.

As a further check on self consistency of Eq. (1.1.142) we can calculate the commutator,

$$\begin{aligned}&\left[SA^{(1)}S^{-1}, SA^{(2)}S^{-1}\right]\\&= SA^{(1)}S^{-1}\, SA^{(2)}S^{-1} - SA^{(2)}S^{-1}\, SA^{(1)}S^{-1},\end{aligned} \tag{1.1.150}$$

where

$$A^{(1)} = A_X^{(1)}\begin{bmatrix}0 & 0 & 0\\0 & 0 & -i\\0 & i & 0\end{bmatrix} + A_Y^{(1)}\begin{bmatrix}0 & 0 & i\\0 & 0 & -0\\-i & 0 & 0\end{bmatrix}, \tag{1.1.151}$$

and

$$\begin{aligned}&SA^{(1)}S^{-1} = SA_X^{(1)}S^{-1} + SA_Y^{(1)}S^{-1}\\&= \frac{A^{(0)}}{\sqrt{2}}e^{i\phi}\begin{bmatrix}0 & 0 & -\sin\alpha+\cos\alpha\\0 & 0 & -i\cos\alpha-\sin\alpha\\i\sin\alpha-\cos\alpha & i\cos\alpha+\sin\alpha & 0\end{bmatrix}.\end{aligned} \tag{1.1.152}$$

Similarly,

$$SA^{(2)}S^{-1} = SA_X^{(1)}S^{-1} + SA_Y^{(2)}S^{-1}$$

$$= \frac{A^{(0)}}{\sqrt{2}} e^{-i\phi} \begin{bmatrix} 0 & 0 & -\sin\alpha - \cos\alpha \\ 0 & 0 & -i\cos\alpha + \sin\alpha \\ i\sin\alpha + \cos\alpha & i\cos\alpha - \sin\alpha & 0 \end{bmatrix} . \tag{1.1.153}$$

Straightforward algebra then shows that,

$$\left[SA^{(1)}S^{-1},\, SA^{(2)}S^{-1}\right] = -A^{(0)2} J_Z , \tag{1.1.154}$$

or in vector notation, we obtain the self consistent result,

$$\boldsymbol{B}^{(3)} = -i \frac{\kappa}{A^{(0)}} \boldsymbol{A}^{(1)} \times \boldsymbol{A}^{(2)} , \tag{1.1.155}$$

which again shows that the cross product of two polar vectors, $\boldsymbol{A}^{(1)}$ and $\boldsymbol{A}^{(2)}$, is the axial vector $\boldsymbol{B}^{(3)}$. (Recall that the cross product of two polar or of two axial vectors both give rise to an axial vector, not to a polar vector [1—4].) Therefore the longitudinal polar vector potential A_Z can be a component of the overall potential four-vector, but cannot be generated by the cross product $\boldsymbol{A}^{(1)} \times \boldsymbol{A}^{(2)}$. The latter always generates an axial vector proportional to $\boldsymbol{B}^{(3)}$.

Now evaluate the commutator $\left[A^{(1)},\, A^{(2)}\right]$, with,

$$A^{(1)} = \frac{A^{(0)}}{\sqrt{2}} e^{i\phi} \begin{bmatrix} 0 & 0 & 1 \\ 0 & 0 & -i \\ -1 & i & 0 \end{bmatrix}, \qquad A^{(2)} = \frac{A^{(0)}}{\sqrt{2}} e^{-i\phi} \begin{bmatrix} 0 & 0 & -1 \\ 0 & 0 & -i \\ 1 & i & 0 \end{bmatrix}; \tag{1.1.156}$$

to obtain

$$\left[A^{(1)}, A^{(2)}\right] = -A^{(0)2} J_Z, \tag{1.1.157}$$

and so,

$$\left[A^{(1)}, A^{(2)}\right] = \left[SA^{(1)}S^{-1}, SA^{(2)}S^{-1}\right]. \tag{1.1.158}$$

This result means that an *O(3)* gauge transformation defined as a Z axis rotation changes $\boldsymbol{A}^{(1)}$ and $\boldsymbol{A}^{(2)}$ but leaves $\boldsymbol{A}^{(1)} \times \boldsymbol{A}^{(2)}$ unchanged. This is an obvious and simple geometrical result which is physically meaningful as a geometric rotation in three dimensions, and which is self consistent with the invariance of $\boldsymbol{B}^{(3)}$ under such a gauge transformation. These concepts do not exist in *U(1)*.

1.3.7 The Topological or Inhomogeneous Term: The Optical Aharonov-Bohm Effect and Topological Phase Effect in *O(3)*

The complete gauge transformation is,

$$\left.\begin{aligned} A^{(1)} &\rightarrow SA^{(1)}S^{-1} - \frac{i}{g}\left(\partial_\mu SS^{-1}\right)^{(1)}, \\ A^{(2)} &\rightarrow SA^{(2)}S^{-1} + \frac{i}{g}\left(\partial_\mu SS^{-1}\right)^{(2)}, \end{aligned}\right\} \tag{1.1.159}$$

and for a *Z* axis rotation,

$$\left[A^{(1)}, A^{(2)}\right] = \left[SA^{(1)}S^{-1}, SA^{(2)}S^{-1}\right], \tag{1.1.160}$$

$$-\frac{i}{g}\left(\partial_\mu SS^{-1}\right)^{(1)} = \frac{1}{g}\partial_\mu \alpha \begin{bmatrix} 0 & -i & 0 \\ i & 0 & 0 \\ 0 & 0 & 0 \end{bmatrix} = \frac{J_Z}{g}\partial_\mu \alpha \, , \tag{1.1.161}$$

$$\frac{i}{g}\left(\partial_\mu SS^{-1}\right)^{(2)} = \frac{1}{g}\partial_\mu \alpha \begin{bmatrix} 0 & i & 0 \\ -i & 0 & 0 \\ 0 & 0 & 0 \end{bmatrix} = -\frac{J_Z}{g}\partial_\mu \alpha \, , \tag{1.1.162}$$

From Eq. (1.1.160) we know that the sum generated by the commutator of inhomogeneous terms and cross terms on the right hand side must be zero. The commutator of inhomogeneous terms is indeed zero,

$$-\frac{1}{g^2}(\partial_\mu \alpha)(\partial_\mu \alpha)^* \left(\begin{bmatrix} 0 & 1 & 0 \\ -1 & 0 & 0 \\ 0 & 0 & 0 \end{bmatrix} \begin{bmatrix} 0 & 1 & 0 \\ -1 & 0 & 0 \\ 0 & 0 & 0 \end{bmatrix} - \begin{bmatrix} 0 & 1 & 0 \\ -1 & 0 & 0 \\ 0 & 0 & 0 \end{bmatrix} \begin{bmatrix} 0 & 1 & 0 \\ -1 & 0 & 0 \\ 0 & 0 & 0 \end{bmatrix} \right) = 0 \, . \tag{1.1.163}$$

Therefore the sum of cross terms must be zero,

$$\begin{aligned} &\left(SA^{(1)}S^{-1}\right)\left(\partial_\mu SS^{-1}\right)^{(2)} - \left(\partial_\mu SS^{-1}\right)^{(1)}\left(SA^{(2)}S^{-1}\right) \\ &+ \left(SA^{(2)}S^{-1}\right)\left(\partial_\mu SS^{-1}\right)^{(1)} - \left(\partial_\mu SS^{-1}\right)^{(2)}\left(SA^{(1)}S^{-1}\right) = 0 \end{aligned} \tag{1.1.164}$$

After some elementary algebra the result reduces to

$$e^{i\phi}\begin{bmatrix} 0 & 0 & ie^{-i\alpha} \\ 0 & 0 & e^{-i\alpha} \\ -ie^{-i\alpha} & -e^{-i\alpha} & 0 \end{bmatrix} + e^{-i\phi}\begin{bmatrix} 0 & 0 & ie^{i\alpha} \\ 0 & 0 & -e^{i\alpha} \\ -ie^{i\alpha} & e^{i\alpha} & 0 \end{bmatrix} = 0\,, \tag{1.1.165}$$

i.e.,

$$e^{i(\phi-\alpha)} = -e^{-i(\phi-\alpha)}\,, \tag{1.1.165a}$$

or,

$$\cos(\phi-\alpha) = 0\,, \tag{1.1.165b}$$

$$\phi \rightarrow \alpha \pm (2n+1)\frac{\pi}{2}\,, \tag{1.1.165c}$$

Therefore the *O(3)* gauge transformation produces a topologically induced change in the electromagnetic phase. A rotation through the angle produces a change $\alpha \pm (2n+1)\pi/2$ in the phase. This is also a polarization change because for instance,

$$(\boldsymbol{i} + i\boldsymbol{j})\, e^{i\phi} \rightarrow (\boldsymbol{i} + i\boldsymbol{j})\, e^{i(\alpha \pm (2n+1)\pi/2)}\,, \tag{1.1.166}$$

and using the angle formulae,

$$\left.\begin{aligned} \cos(A \pm B) &= \cos A \cos B \mp \sin A \sin B\,, \\ \sin(A \pm B) &= \sin A \cos B \pm \cos A \sin B\,, \end{aligned}\right\} \tag{1.1.167}$$

it follows that

$$\left.\begin{aligned} Re\left((\boldsymbol{i}+i\boldsymbol{j})e^{i\phi}\right) &= \cos\phi\,\boldsymbol{i} - \sin\phi\,\boldsymbol{j} \\ &\rightarrow \pm\left(\sin\phi_0\boldsymbol{i} + \cos\phi_0\boldsymbol{j}\right), \end{aligned}\right\} \tag{1.1.168}$$

where $\phi_0 = \alpha$.

This result clearly shares the features of the topological phase effect, for example, winding an optical fiber on a drum, and sending a linearly polarized laser beam through it produces a rotation of the linear polarization plane [8,9]. This is in *O(3)* an optical Aharonov Bohm effect as argued. In *U(1)* the same effect is random, and unphysical. This seems to be further clear empirical reason for preferring *O(3)* to *U(1)* and the observation of the topological phase in this manner is also an observation of the optical Aharonov-Bohm effect. For example, a rotation of $3\pi/2$ increases ϕ in Eq. (1.1.165) by the same amount, $3\pi/2$, and changes the *polarization* of the light beam. For example, for $n = 0$,

$$\left.\begin{aligned} \sin\left(\alpha + \frac{\pi}{2}\right) &= \cos\alpha \qquad (\neq \sin\alpha \text{ in general}), \\ \cos\left(\alpha + \frac{\pi}{2}\right) &= -\sin\alpha \qquad (\neq \cos\alpha \text{ in general}). \end{aligned}\right\} \tag{1.1.169}$$

Since gauge transformation in *O(3)* is a physical (or geometrical) rotation, the rotation of the direction of the light beam as it propagates through an optical fiber wound about the *Z* axis as a helix [8,9] is a geometrical process that is a gauge transformation, one which can be observed empirically to change the polarization of that light beam, *QED*. The geometrical details are different, because the helical rotation of a beam propagating within a fiber is not the same as a straightforward rotation of that beam about its own propagation axis *Z* when the latter is held constant, but the overall result is the same, a change of polarization of the light beam. Such a phenomenon has no existence in *U(1)*.

References

[1] M. W. Evans and J.-P. Vigier, *The Enigmatic Photon, Vol. 1: The Field* $\boldsymbol{B}^{(3)}$ (Kluwer Academic, Dordrecht, 1995).
[2] M. W. Evans and J.-P. Vigier, *The Enigmatic Photon, Vol. 2: Non-Abelian Electrodynamics* (Kluwer Academic, Dordrecht, 1995).
[3] M. W. Evans , J.-P. Vigier, S. Roy, and S. Jeffers, *The Enigmatic Photon, Vol. 3: Theory and Practice of the* $\boldsymbol{B}^{(3)}$ *Field* (Kluwer Academic, Dordrecht, 1996).
[4] M. W. Evans, J.-P. Vigier, and S.Roy, *The Enigmatic Photon, Vol. 4: New Directions* (Kluwer Academic, Dordrecht, 1998).
[5] L. H. Ryder, *Quantum Field Theory* (Cambridge University Press, Cambridge, 1987).
[6] M. W. Evans, *Physica B*, **182**, 227 (1992).
[7] W.K.H. Panofsky and M. Phillips, *Classical Electricity and Magnetism* (Addison-Wesley, Reading, 1962).
[8] T.W. Barrett and D. M. Grimes, eds., *Advanced Electromagnetism, Foundations, Theory and Applications*, Chap. 1 (World Scientific, Singapore, 1995).
[9] T. W. Barrett in A. Lakhtakia, ed. *Essays on the Formal Aspects of Electromagnetic Theory*, (World Scientific, Singapore, 1993).
[10] R. Aldrovandi, Ref. 8, pp. 3 ff.
[11] T. E. Bearden, personal communications.
[12] R. S. Fritzius, Critical Researches on General Electrodynamics *Apeiron*, 1998.
[13] W. Ritz, *Ann. Chim. Phys.*, **13**, 145 (1908).
[14] M. W. Evans in I. Prigogine and S. A. Rice, eds., *Advances in Chemical Physics*, Vol. 81, pp 361—702 (Wiley, New York, 19 92).
[15] B. L. Silver, *Irreducible Tensor Theory* (Academic, New York, 1976).
[16] A. O. Barut, Electrodynamics and Classical Theory of Fields and Particles (MacMillan, New York, 1964).
[17] D. Corson and P. Lorrain, *Introduction to Electromagnetic Fields and Waves* (W. H. Freeman & Co., San Francisco, 1962).

[18] M. W. Evans and S. Kielich eds., *Modern Nonlinear Optics*, Vols. 85(1), 85(2), 85(3) of Advances in Chemical Physics, I. Prigogine and S. A. Rice, eds. (Wiley Interscience, New York, 1993/1994/1997).

[19] J. D. Jackson, *Classical Electrodynamics* (Wiley, New York, 1962).

[20] Definitions in Ref. 10.

Chapter 2

The Geometry of Gauge Fields

Contemporary bundle tangent theory is able to establish the basic structure of any gauge theory from pure geometry. It can be shown [1] that the internal space is a symmetry space. Vector fields, forms and tensors on the basic manifold are related to their correspondents on the bundle. Vector fields are lifted by a section to certain fields on the bundle and this is pure geometry as is well known in contemporary mathematical physics. A frame $\left[e_\mu\right]$ on the basic space will be taken by a section σ into a set of basic fields $X_\mu = \sigma\left(e_\mu\right)$. Around any point of the bundle there exists a *separated* basis, called a direct product basis, formed by the basic fields x_μ and the fundamental fields X_a. In this (direct product) basis the commutation relations are [1],

$$\left[X_\mu, X_\nu\right] = C^\lambda_{\mu\nu} X_\lambda , \tag{1.2.1}$$

$$\left[X_\mu, X_a\right] = 0 , \tag{1.2.2}$$

$$\left[X_a, X_b\right] = f^c_{ab} X_c , \tag{1.2.3}$$

As described in Ref. 1, Eq. (1.2.2) establishes the independence of the algebra of the fundamental and basic fields.

2.1 Application to Electromagnetic Theory with Internal Space ((1), (2), (3))

Usually, electromagnetism is described as a gauge theory with *U(1)* internal symmetry [2—4]. However, in Chap. 1 we have developed a gauge theory of electromagnetism with an internal space ((1), (2), (3)) which is a physical space with *O(3)* symmetry. Does this gauge theory comply with Eqs. (1.2.1) to (1.2.3)? This question can be tested with a particular choice of generators for the X_μ fields. If, for example, we choose the X_μ to be rotation generators of the Lorentz group we obtain [5—8] for Eq. (1.2.1),

$$\left.\begin{aligned} \left[J_1, J_2\right] &= iJ_3 , \\ \left[J_2, J_3\right] &= iJ_1 , \\ \left[J_3, J_1\right] &= iJ_2 \end{aligned}\right\} \tag{1.2.4}$$

so $f^c_{ab} = i,\ a = X,\ b = Y,\ c = Z$. Similarly for Eq. (1.2.3),

$$\left[J^{(1)}, J^{(2)}\right] = -J^{(3)*} , \tag{1.2.5}$$

et cyclicum, so $f^c_{ab} = -1,\ a = (1),\ b = (2),\ c = (3)$. However, when we come to test Eq. (1.2.2), we obtain results such as

$$\left[J_1, J^{(2)}\right] = \frac{1}{\sqrt{2}} J^{(3)*} , \tag{1.2.6}$$

and so forth. This simply means that the internal gauge space as used in Chap. 1 is not independent of space-time, as in Eqs. (1.2.1), (1.2.2), and (1.2.3). How are we to justify the gauge theory of Chap. 1 therefore in the general structure described for example by Aldrovandi [1]? It turns out that the answer is to be found in the theory of extended Lie algebra.

Note firstly that the X^μ vector field symbols used by Aldrovandi [1] can represent space-time translation generators, for example, or rotation generators. In a simple or direct product Lie algebra he proves that all gauge theories must have the following structure (his Eqs. (61) to (63)) [1],

$$\left[X'_\mu, X'_\nu\right] = C^\lambda_{\mu\nu} X'_\lambda - F^a_{\mu\nu} X_a , \tag{1.2.7}$$

$$\left[X'_\mu, X_a\right] = 0 , \tag{1.2.8}$$

$$\left[X_a, X_b\right] = f^c_{ab} X_c , \tag{1.2.9}$$

where $X'_\mu = X_\mu - A^a_\mu X_a$. If X_μ is a translation generator proportional to ∂_μ [4], then X'_μ is a covariant derivative. The X_a generators can be matrix generators for example.

In simple gauge theories such as these, Eq. (1.2.8) shows that the internal (gauge) space and the external space are independent. The connection A^a_μ modifies the translation generator $X_\mu := \partial_\mu$ and alters space-time homogeneity [1]. This is a key point in all gauge theories applied to electrodynamics. Essentially, the non-linearities in electrodynamics become a property of space-time in the spirit of general relativity. The *O(3)* symmetry electrodynamics is only one example of many possible self consistent theories of non-linear optics, all of which reduce to the Maxwellian formalism in the linear limit. In a simple gauge theory such as the Yang-Mills theory the underlying group Jacobi identities completely determine the structure of the theory, which is fixed by Eqs. (1.2.7) to (1.2.9). The Bianchi identity for example is formed from the Jacobi identity through the use of covariant derivatives [1], and all of these results are purely geometrical. They are of the type used in particle physics, where it is usually assumed that the internal space is independent of Minkowski space-time.

2.1.1 Extended Lie Algebra

A rigorous geometrical basis for the theory used in Chap. 1 can be given using a simple example of the Lie algebra extension theory given by Aldrovandi [2.8] in his section 6.1. This is developed in this section using the Lie algebra of rotation generators in the basis (X, Y, Z) and the basis ((1), (2), (3)). The L algebra [1] is defined by

$$\left[J_X, J_Y\right]_L = iJ_Z, \tag{1.2.10}$$

and the V algebra by:

$$\left[J^{(1)}, J^{(2)}\right]_V = -J^{(3)*}. \tag{1.2.11}$$

Given a Lie algebra L and a representation ρ of L on another algebra V we produce a joint algebra E encompassing L and V, following the methods given by Aldrovandi [1]. The algebra E is an extension of L by V through ρ. The extension of V to E is an inclusion such that

$$\left[J^{(1)}, J^{(2)}\right]_V = \left[J^{(1)}, J^{(2)}\right]_E = f_{(1)(2)}^{(3)*} J^{(3)*}, \tag{1.2.12}$$

so, $f_{(1)(2)}^{(3)*} = -1$. The extension of L to E is a mapping such that,

$$\left.\begin{aligned} \sigma &: L \rightarrow E, \\ \sigma &: J_\mu \rightarrow J^{(a)}; \qquad (a) = (1),\ (2),\ (3), \end{aligned}\right\} \tag{1.2.13}$$

and,

$$\left[J^{(1)}, J^{(2)}\right]_E = iC_{(1)(2)}^{(3)*} J^{(3)*}, \tag{1.2.14}$$

where

$$\left[J_\mu, J_\nu\right] = C_{\mu\nu}^{\rho} J_\rho. \tag{1.2.15}$$

Therefore $C_{(1)(2)}^{(3)*} = i$ and,

$$iC_{(1)(2)}^{(3)*} = f_{(1)(2)}^{(3)*} \,. \tag{1.2.16}$$

The mapping (1.2.13) means that the Cartesian space of J_μ in L is extended to a complex spherical space in E.

The Lie algebras L and V have been combined into a Lie algebra E with an underlying vector space $L \oplus V$, the direct *sum* of those of L and V. In general, L and V can be combined [1] to give many different extended algebras E. In this case E is an algebra that incorporates the Cartesian basis (L) and the spherical basis V where L describes a Cartesian basis and V a spherical basis only. For rotation generators, the extended Lie algebra E is given by

$$\left[J^{(1)}, J^{(2)}\right]_E = \left[J^{(1)}, J^{(2)}\right]_V = f_{(1)(2)}^{(3)*} J^{(3)*} , \tag{1.2.17}$$

$$\left[J^{(1)}, J^{(2)}\right]_E = \rho\left(J_\mu\right) J^{(2)} = iC_{(1)(2)}^{(3)*} J^{(3)*} , \tag{1.2.18}$$

$$\left[J_\mu, J_\nu\right]_E = C_{\mu\nu}^{\rho} J_\rho - \beta_{\mu\nu}^{(3)*} J^{(3)*} , \tag{1.2.19}$$

where the constants $\beta_{\mu\nu}^{(3)*}$ measure the departure from homomorphism [1].

2.1.2 Extended Lie Algebra with Connections

Equation (1.2.18) above measures that the coupling between the internal space ((1), (2), (3)) and the extended space of the E Lie algebra. This can be made clear by writing the commutator on the left hand side of Eq. (1.2.18) as [5—8],

$$\left[J^{(1)}, J^{(2)}\right]_E = \frac{1}{i}\left[\frac{1}{\sqrt{2}}\left(J_X - iJ_Y\right), J^{(2)}\right]. \tag{1.2.20}$$

This means that the theory developed in Chap. 1 is an extended Lie algebra with connections, which is described in general gauge theory by Aldrovandi on his page 39 [1], his Eqs. (105) to (111). In this extended Lie algebra, the connection is denoted [1] B_μ^a and

$$X_\mu^{/} = X_\mu - B_\mu^a X_a . \tag{1.2.21}$$

The commutator relations become

$$\left.\begin{aligned} \left[X_\mu^{/}, X_\nu^{/}\right] &= C_{\mu\nu}^{\rho} X_\rho^{/} - \beta_{\mu\nu}^{/c} X_c , \\ \left[X_\mu^{/}, X_b\right] &= C_{\mu b}^{/c} X_c , \\ \left[X_a, X_b\right] &= f_{ab}^{c} X_c , \end{aligned}\right\} \tag{1.2.22}$$

where [1],

$$\left.\begin{aligned} \beta_{\mu\nu}^{/c} &= \beta_{\mu\nu}^{c} + K_{\mu\nu}^{c} , \\ K_{\mu\nu}^{c} &= C_{\mu a}^{c} B_\nu^a - C_{\nu a}^{c} B_\mu^a - B_\rho^c C_{\mu\nu}^{\rho} - f_{ab}^{c} B_\mu^a B_\nu^b \\ C_{\mu b}^{/c} &= C_{\mu b}^{c} - B_\mu^a f_{ab}^{c} . \end{aligned}\right\} \tag{1.2.23}$$

If $C_{\mu b}^{/c} = \beta_{\mu\nu}^{/c} = 0$ there is no extension [1].

We are now in a position to check this extended gauge theoretical structure against the *O(3)* gauge theory given by Ryder [4] in his Chap. 3. This theory is in turn the basis of our development in Chap. 1. We shall first show that even *U(1)* electrodynamics, seen as a gauge theory, is a special case of Eq. (1.2.22), but with $C^{/c}_{\mu b}$ *not* equal to zero. This means that the Lie algebra underlying ordinary *U(1)* electrodynamics is an *extended* Lie algebra, and so the spaces *L* and *V* are *not* independent, even in Maxwellian electrodynamics seen as a gauge field theory. So application of gauge theory, with affine algebra, to electrodynamics is different in principle from its application to elementary particle physics, if- when the latter takes the two spaces to be independent. The only idea in common is that space-time is made inhomogeneous.

The special case of *U(1)* electromagnetism can be recovered from Eqs. (1.2.22) as follows. Firstly define the covariant derivatives by taking $X^{/}_{\mu}$ and $X^{/}_{\nu}$ to be extended translation generators,

$$X^{/}_{\nu} = \partial_{\mu} - igA_{\mu}, \qquad X^{/}_{\nu} = \partial_{\nu} - igA_{\nu}, \tag{1.2.24}$$

where $g = e$, the charge on the proton, and A_{μ} and A_{ν} are the *U(1)* four potentials [4]. Then,

$$\left[X^{/}_{\mu}, X^{/}_{\nu}\right] = -ie\left(\partial_{\mu} A_{\nu} - \partial_{\nu} A_{\mu}\right) = -ieF_{\mu\nu}, \tag{1.2.25}$$

where $F_{\mu\nu}$ is the ordinary *U(1)* field tensor. From Aldrovandi's Eq. (99),

$$\left[X_{\mu}, X_{\nu}\right]_{E} = C^{\rho}_{\mu\nu} X_{\rho} = 0, \tag{1.2.26}$$

because $X_{\mu} = \partial_{\mu}$; $X_{\nu} = \partial_{\nu}$; $X_{\rho} = \partial_{\rho}$ are translation generators within a proportionality factor [4]. Therefore,

$$C^{\rho}_{\mu\nu} = 0, \tag{1.2.27}$$

and

$$\beta^{/c}_{\mu\nu} = igF_{\mu\nu}. \tag{1.2.28}$$

Since we are dealing with an affine algebra, we obtain in Aldrovandi's Eq. (100),

$$\beta^{c}_{\mu\nu} = 0 \,. \tag{1.2.29}$$

Furthermore, in *U(1)*,

$$X_a = X_b = X_c = 1 \,, \tag{1.2.30}$$

being interpreted as rotation generators of *U(1)* [4]. Therefore,

$$f^{c}_{ab} = 0 \,, \tag{1.2.31}$$

and

$$\beta'^{c}_{\mu\nu} = K^{c}_{\mu\nu} = C^{c}_{\mu a} B^{a}_{\nu} - C^{c}_{\nu a} B^{a}_{\mu} \,. \tag{1.2.32}$$

We can identify:

$$\left.\begin{aligned} C^{c}_{\mu a} &:= ig\partial_{\mu} \,, & C^{c}_{\nu a} &:= ig\partial_{\nu} \,, \\ B^{a}_{\nu} &:= A_{\nu} \,, & B^{a}_{\mu} &:= A_{\mu} \,. \end{aligned}\right\} \tag{1.2.33}$$

As described by Aldrovandi the *C's* are interpreted as matrix operators which in *U(1)* are 1×1 matrices, *e.g.*,

$$C^{b}_{\mu c} = \left(X_{\mu}\right)^{b}_{c} = \partial_{\mu} \quad \text{etc.}\,. \tag{1.2.34}$$

When we come to examine the commutator (1.2.22b) we find,

$$\left[X'_{\mu}, X_b\right] = \left[\partial_{\mu} - igA_{\mu}, 1\right] = C'^{c}_{\mu b} X_c = 0 \,. \tag{1.2.35}$$

We have shown that $f_{ab}^{c} = 0$, so,

$$C_{\mu b}^{/c} = C_{\mu b}^{c} = \partial_{\mu}, \tag{1.2.36}$$

and

$$C_{\mu b}^{/c} X_{c} = \partial_{\mu} 1 = 0. \tag{1.2.37}$$

So Eq. (1.2.35) becomes

$$0 = C_{\mu b}^{c} 0, \qquad C_{\mu b}^{c} \neq 0, \tag{1.2.38}$$

and since the spaces L and V decouple if and only if $C_{\mu b}^{c} = 0$, *the U(1) theory of electrodynamics is a gauge theory in an extended Lie algebra.* So the fundamental and basic fields in *U(1)* electrodynamics occur in an E space; $E = L \oplus V$, the direct sum of L and V. The internal *U(1)* gauge space is *not* independent of the space-time of the gauge theory. This is of course a physical result, because the potential four-vector A_{μ} of the *U(1)* theory is well defined in both L and V.

These points appear not to have been realized hitherto, or not made clear. The application as exemplified by Ryder, of gauge field theory to classical electrodynamics implies an extended Lie algebra whose two spaces are *not* independent. If elementary particle theory is to be defined as a gauge field theory then it is usually assumed that the two spaces are independent. This hypothesis is justified by its success in particle physics, but evidently, the photon does not fit into a gauge field theory whose spaces are independent, even in the *U(1)* linear approximation. This questions the standard model again at a fundamental level, and questions the assertion that the photon is a particle, or at least the same type of particle as for example a quark or electron. This is a rigorous result of pure geometry applied as we have just demonstrated to *U(1)* electrodynamics, the kind of electrodynamics that is usually quantized to give the photon.

On the classical level, the same geometrical methods of advanced fiber bundle theory show [1] that there is no conceptual problem whatsoever in replacing *U(1)* by *O(3)* in classical electrodynamics, we are simply changing the symmetry of an internal space which by definition is a symmetry space if we are to apply group theoretic restrictions to the covariant derivative. (More generally we can lift these restrictions and make the covariant derivative a Taylor series for example.) Extended gauge theory [1] gives as rigorous a basis for *O(3)* as it does for *U(1)*, or any other group theoretic restriction on the covariant derivative. In the last analysis, such a description is a guess about the vacuum, or in-homogeneity of space-time, and shifts the description of what mediates interaction between two charges from the field to the potential and to space-time itself, in the spirit of general relativity. In this scenario, the classical electromagnetic field is the result of a round trip with covariant derivatives [4]. If the round trip has a physical effect, the field is not zero. The *U(1)* hypothesis makes the covariant derivative linear in the potential four-vector. A round trip produces the familiar four-curl, but from a theory akin to *general* relativity [4] in which space-time itself is given a structure. This simple linear hypothesis results in Maxwell's equations. The next simplest guess, or hypothesis, is *O(3)* electrodynamics, in which the covariant derivative contains rotation generators of *O(3)*, and in which the field is non-linear in the potential as developed in Chap 1. The *O(3)* guess results in a theory which is already much richer than *U(1)*, but is still a simple guess. Proceeding in this way there emerges a set of classical electrodynamic theories, each member of which is as rigorous as *U(1)*. The differences between each member of this set of theories show up most vividly in the vacuum. For example, as we have seen in Chap. 1, *O(3)* gives vacuum polarization and magnetization (the B cyclic theorem), which are all missing from *U(1)*. Similarly, an *SU(3)* group theoretic guess will bring out a far richer structure than *O(3)* and so on. In this way we can begin to describe the various non linear optical phenomena [5—8] which have no existence at all in *U(1)*. The conventional [9] phenomenological approach simply inserts non-linear terms into *U(1)* through the classical constitutive relations: a hybrid, self-contradictory, approach. (Recall, for example, that $\boldsymbol{A}^{(1)} \times \boldsymbol{A}^{(2)}$ has no existence in *U(1)*, but is thrown ad hoc into the theory in order to be able to describe the inverse Faraday effect.)

The *O(3)* theory emerges from the general geometrical equations if the internal space V has this symmetry, so that X_a, X_b and X_c become rotation generators of *O(3)*,

$$\left.\begin{aligned} X_a := J_a, \quad X_b := J_b, \quad X_c := J_c, \\ \left[J_a, J_b\right] = f_{ab}^{c} J_c, \quad f_{ab}^{c} \neq 0. \end{aligned}\right\} \tag{1.2.39}$$

The *O(3)* covariant derivatives are extended space-time translation generators of the general theory given by Aldrovandi [1],

$$\left.\begin{aligned} X_\mu' = \partial_\mu - igA_\mu^a J_a := \partial_\mu - igA_\mu := D_\mu, \\ X_\nu' = \partial_\nu - igA_\nu^a J_a := \partial_\nu - igA_\nu := D_\nu, \end{aligned}\right\} \tag{1.2.40}$$

and g is proportional to the elementary charge e through different coefficients in free space and in the presence of matter. This has no effect on the *O(3)* symmetry of the internal gauge space. So [5—8],

$$\left[X_\mu', X_\nu'\right]_E = -ig\left(\partial_\mu A_\nu - \partial_\nu A_\mu - ig\left[A_\mu, A_\nu\right]\right), \tag{1.2.41}$$

and the commutator $\left[A_\mu, A_\nu\right]$ is non-zero, making the theory non-linear.

Note that this is still a theory of electromagnetism, the elementary charge e still appears in it, and the potentials are electromagnetic potentials. The theory is, in the last analysis, a purely geometrical description of classical electromagnetism.

As in *U(1)* theory,

$$C_{\mu\nu}^{\rho} = 0, \tag{1.2.42}$$

because unextended translation generators commute, *e.g.* $\left[\partial_\mu , \partial_\nu\right] = 0$ as used for example by Ryder [4] in an *SU(2)* symmetry gauge field theory of elementary particles. So,

$$\left[X'_\mu , X'_\nu\right] = -\beta'^c_{\mu\nu} X_c , \tag{1.2.43}$$

indicating the existence of an internal vector space as in the notation of Eqs. (1.1.38) of Chap. 1; and so, as in Eq. (3.169) of Ryder [4],

$$\beta'^c_{\mu\nu} = ig\left(\partial_\mu A^c_\nu - \partial_\nu A^c_\mu - ig\epsilon_{cab} A^a_\mu A^b_\nu\right) . \tag{1.2.44}$$

If we assume no departure from homomorphism [1], *i.e.*, that,

$$\beta^c_{\mu\nu} = 0 , \tag{1.2.45}$$

we obtain,

$$\beta'^c_{\mu\nu} = C^c_{\mu a}\beta^a_\nu - C^c_{\nu a}\beta^a_\mu - f^c_{ab} B^a_\mu B^b_\nu , \tag{1.2.46}$$

and so,

$$\left.\begin{array}{c} C^c_{\mu a} B^a_\nu := ig\partial_\mu A^c_\nu , \qquad C^c_{\nu a} B^a_\mu := ig\partial_\nu A^c_\mu \\ igA^a_\mu := B^a_\mu , \qquad igA^a_\nu := B^a_\nu , \\ f^c_{ab} = i\epsilon_{cab} . \end{array}\right\} \tag{1.2.47}$$

The spaces L and V are connected into an extended Lie algebra because

$$\begin{aligned}\left[X_\mu^{/}, X_b\right]_E &= \left[\partial_\mu - igA_\mu, J_b\right] = \left[\partial_\mu, J_b\right] - ig\left[A_\mu, J_b\right] \\ &= \partial_\mu J_b - igA_\mu^a\left[J_a, J_b\right] \\ &= \partial_\mu J_b - igA_\mu^a f_{ab}^c J_c .\end{aligned} \tag{1.2.48}$$

This result, in Aldrovandi's notation, is identified as,

$$\begin{aligned} C_{\mu b}^c X_c &:= \partial_\mu J_b = 0 , \\ B_\mu^a f_{ab}^c X_c &:= igA_\mu^a f_{ab}^c J_c \end{aligned} . \tag{1.2.49}$$

The *O(3)* theory of classical electromagnetism is therefore an example of a gauge field theory in an affine space with $E = L \oplus V$. So all the development by Aldrovandi in his pages 39 ff. can be taken over unchanged as a description of *O(3)* electrodynamics. This means that all the insights on pp. 39 ff. of Aldrovandi [1] can be implemented, including those in unified field theory. The result is a powerful support for *O(3)* electrodynamics based on pure geometry. No physics has yet entered the scene [1]. In other words we have guessed that space-time can be made inhomogeneous by the imposition of an internal *O(3)* symmetry in a gauge field theory. Metaphorically, Maxwell guessed that this symmetry is *U(1)*. (Historically, gauge field theories were not, of course, available to him.)

2.2 The Geometrical Meaning of *O(3)* Electrodynamics

The results of Sec. 2.1 mean that *O(3)* electrodynamics is *completely* defined in contemporary geometrical theories, provided that $E = L \oplus V$, *i.e.*, that E is the direct sum of L and V. For example, *O(3)* electrodynamics can be fully developed using exterior derivatives in an anholonomic basis, extending the Maurer-Cartan equations as described in Aldrovandi's Sec. 6.2. The *O(3)* electrodynamics can also be developed as a field algebra on manifolds, leading to a similarity with gravitational theories as developed in Vol. 4 of this series [8]. The key empirical difference between *U(1)* and *O(3)* electrodynamics is that the commutator $\left[A_\mu , A_\nu\right]$ is non zero in *O(3)*, as observed in the inverse Faraday effect of nonlinear magneto-optics [9].

2.3 The Field Equations of *O(3)* Electrodynamics

Physics enters the scene when we come to consider field equations [1]. In this section, their complete self-consistency in classical *O(3)* electrodynamics is demonstrated for the free field and in the presence of field matter interaction.

2.3.1 The *O(3)* Field Equations in Free Space

It is argued in this section that the following *O(3)* free space field equations are rigorously self-consistent,

$$D_\nu \tilde{\boldsymbol{G}}^{\mu\nu} := \boldsymbol{0}, \tag{1.2.50}$$

$$D_\nu \boldsymbol{G}^{\mu\nu} = \frac{\boldsymbol{J}\left(\text{vac}\right)}{\epsilon_0}, \tag{1.2.51}$$

where $\tilde{\boldsymbol{G}}^{\mu\nu}$ is the dual of the *O(3)* field tensor defined in Chap. 1,

$$\tilde{\boldsymbol{G}}^{\mu\nu} := \frac{1}{2}\epsilon^{\mu\nu\rho\sigma}\boldsymbol{G}_{\rho\sigma}, \tag{1.2.52}$$

and where $\boldsymbol{J}^\mu\left(\text{vac}\right)$ is a vacuum Noether current, or helicity current, to be defined. Eq. (1.2.50) is the Feynman-Jacobi identity for an *O(3)* symmetry gauge field theory [4], one in which the covariant derivative is defined in terms of *O(3)* rotation generators. The *O(3)* field tensor $\boldsymbol{G}^{\mu\nu}$ is defined and discussed in detail in Chap. 1. The equations (1.2.50) and (1.2.51) are therefore,

$$\partial_\nu \tilde{\boldsymbol{G}}^{\mu\nu} + g\boldsymbol{A}_\nu \times \tilde{\boldsymbol{G}}^{\mu\nu} = \boldsymbol{0}\,, \tag{1.2.53}$$

$$\partial_\nu \boldsymbol{G}^{\mu\nu} + g\boldsymbol{A}_\nu \times \boldsymbol{G}^{\mu\nu} = \frac{\boldsymbol{J}^\mu(\mathrm{vac})}{\epsilon_0}\,, \tag{1.2.54}$$

for the free classical field. Equations (1.2.53) and (1.2.54) use the same notation as in Ryder's discussion of Yang-Mills theory [8], but as discussed, form an extended Lie algebra. The coefficient g for the free field is,

$$g = \frac{\kappa}{A^{(0)}} = \frac{e}{\hbar}\,, \tag{1.2.55}$$

and is proportional to the elementary charge e after quantization [5—8],

$$e = \hbar\left(\frac{\kappa}{A^{(0)}}\right). \tag{1.2.56}$$

This concept of photon momentum $\hbar\kappa$ occurs after quantization of the *U(1)* theory, but usually, it is not clear that this quantum of momentum, the photon momentum, is equal to $eA^{(0)}$ for the free field. The conceptual problem posed by Eq. (1.2.55) is the presence of e in the free field, and as argued already, its presence does not mean that the field is charged. It means that the field is C negative. Therefore non-Abelian gauge field theory such as *O(3)* classical electrodynamics allows charge quantization, the elementary charge e being that on the proton, minus the charge on the electron. Equation (1.2.56) is similar to Planck quantization, $En = \hbar\omega$, of the energy. The self consistency of this result is illustrated through the fact that an electron accelerated to c becomes the electromagnetic field in free space as described by Jackson [2], and charge conservation means that e is present in the free field. On an elementary physical level, the electromagnetic field must be C negative for one charge to influence another through the field. In action at a distance theories the same must apply, for example in Schwarzschild's delayed action at a distance theory of 1902 [10] the potential is C negative. On the classical level, the factor g in free space is the wavevector magnitude divided by $A^{(0)}$, and so g is C negative as required. In electrostatics, g goes to zero, and the $O(3)$ theory takes on a linear form, giving the Coulomb, Gauss and Ampère Laws of electrostatics and magnetostatics. This is easy to see because g goes to zero gives a linear theory which has the same structure as the familiar Maxwellian theory, except for the presence of indices (1) and (2), indicating complex conjugation. In the static limit however, we can use a real potential four-vector, so that the indices (1) and (2) are equal. (Complex conjugation does not affect a real valued variable.)

More subtly, we must consider whether the elementary magnetic flux density on one photon, which is the elementary magnetic fluxon [8], divided by a quantization volume, is localized or not after quantization. It is well known that the photon, the quantum of energy, is not localized, and that the photon can be created and destroyed with creation and annihilation operators without affecting the principle of conservation of energy. These are features of the quantized $U(1)$ electromagnetic field. After quantization of $O(3)$ however, we find Eq. (1.2.56), and it seems that the fluxon $\hbar/e$ may share these properties of being non localized with the quantized unit of energy, the conventional photon, $\hbar\omega$. It then seems appropriate to ask whether $\hbar/e$ divided by the quantization volume can also be created and destroyed statistically within the quantized $O(3)$ field without violating the principle of conservation of charge. This is not a feature of the quantized $U(1)$ field.

2.3.2 Self Consistency of Eqs. (1.2.50) and (1.2.51)

Equations (1.2.50) and (1.2.51) are self consistent and consistent with the definition of $\boldsymbol{A}_\mu$ and $\boldsymbol{G}^{\mu\nu}$ used in Chap. 1. The detailed proof of this self consistency is given in this section. The solution of the two field equations (1.2.50) and (1.2.51) must be consistent with the fact that the B cyclic theorem is produced from the fundamental definition of the field tensors appearing in the field equations. This can be so if and only if,

$$\boldsymbol{A}_\nu \times \tilde{\boldsymbol{G}}^{\mu\nu} = \boldsymbol{0}\,, \tag{1.2.57}$$

$$\boldsymbol{A}_\nu \times \boldsymbol{G}^{\mu\nu} = \frac{\boldsymbol{J}(\text{vac})}{\epsilon_0}\,, \tag{1.2.58}$$

These conditions give the *O(3)* field equations in the form,

$$\partial_\nu \tilde{\boldsymbol{G}}^{\mu\nu} = \boldsymbol{0} \tag{1.2.59}$$

$$\partial_\nu \boldsymbol{G}^{\mu\nu} = \boldsymbol{0}\,, \tag{1.2.60}$$

where

$$\tilde{\boldsymbol{G}}^{\mu\nu} = \tilde{G}^{\mu\nu(1)}\boldsymbol{e}^{(1)} + \tilde{G}^{\mu\nu(2)}\boldsymbol{e}^{(2)} + \tilde{G}^{\mu\nu(3)}\boldsymbol{e}^{(3)}\,, \tag{1.2.61}$$

$$\boldsymbol{G}^{\mu\nu} = G^{\mu\nu(1)}\boldsymbol{e}^{(1)} + G^{\mu\nu(2)}\boldsymbol{e}^{(2)} + G^{\mu\nu(3)}\boldsymbol{e}^{(3)}\,, \tag{1.2.62}$$

so we obtain the equations for indices (1) and (2),

$$\partial_\nu \tilde{G}^{\mu\nu(1)} = \partial_\nu \tilde{G}^{\mu\nu(2)} = 0\,, \tag{1.2.63}$$

$$\partial_\nu G^{\mu\nu(1)} = \partial_\nu G^{\mu\nu(2)} = 0\,, \tag{1.2.64}$$

and the equations for the $\boldsymbol{B}^{(3)}$ field,

$$\partial_\nu \tilde{G}^{\mu\nu(3)} = \partial_\nu G^{\mu\nu(3)} = 0\,. \tag{1.2.65}$$

Equations (1.2.63) and (1.2.64) are formally identical with the Maxwell equations in free space for the complex field tensor components $G^{\mu\nu(1)} = G^{\mu\nu(2)*}$ and its dual. Equation (1.2.65) is not present in *U(1)* electrodynamics and in vector notation gives the $\boldsymbol{B}^{(3)}$ field equation in free space,

$$\nabla \times \boldsymbol{B}^{(3)} = \boldsymbol{0}\,, \qquad \frac{\partial \boldsymbol{B}^{(3)}}{\partial t} = \boldsymbol{0}\,. \tag{1.2.66}$$

It will be shown that Eq. (1.2.57) produces the B cyclic equations self consistently. Equation (1.2.58) produces the helicity current, which depends on $\boldsymbol{B}^{(3)} \neq \boldsymbol{0}$; $\boldsymbol{A}^{(3)} \neq \boldsymbol{0}$. These concepts are not available in *U(1)* electromagnetism. Equations (1.2.57) and (1.2.58) also produce the vacuum Maxwell equations given, self-consistently, that the plane waves $\boldsymbol{B}^{(1)}$ and $\boldsymbol{B}^{(2)}$ are solutions of the Maxwell equations and that $\boldsymbol{B}^{(3)}$ is a solution of Eqs. (1.2.65) and (1.2.66), being phase free. A third self-consistency check is that $\boldsymbol{B}^{(1)}$, $\boldsymbol{B}^{(2)}$ and $\boldsymbol{B}^{(3)}$ are linked by the B cyclic theorem which is given by Eq. (1.2.57). A fourth check for self consistency is given by the fact that the *O(3)* electrodynamical equations in free space give $\boldsymbol{E}^{(3)} = \boldsymbol{0}$. The $\boldsymbol{B}^{(3)}$ field is not accompanied by an $\boldsymbol{E}^{(3)}$ field [5—8] as shown by Eq. (1.2.57) to (1.2.66), and as shown empirically by Raja *et al.* [8] and Compton *et al.* [8]. There is no Faraday induction due to $\boldsymbol{B}^{(3)}$ and it is observed experimentally through $\boldsymbol{A}^{(1)} \times \boldsymbol{A}^{(2)}$ in field-matter interaction, for example the magnetization of the inverse Faraday effect is due to $\boldsymbol{A}^{(1)} \times \boldsymbol{A}^{(2)}$ as shown in Chap. 1. Note that $\boldsymbol{B}^{(3)}$ is therefore a fundamental field of *O(3)* electrodynamics, but does not occur in *U(1)* electrodynamics. As shown earlier in this chapter, both theories are rigorous gauge field theories in an extended Lie algebra $E = T \oplus V$. However, *O(3)* has several advantages over *U(1)* and gives insights and concepts which *U(1)* does not. For example, the commutator $\boldsymbol{A}^{(1)} \times \boldsymbol{A}^{(2)}$ is zero by definition in *U(1)* electrodynamics, yet $\boldsymbol{A}^{(1)} \times \boldsymbol{A}^{(2)}$ is an empirical observable of the inverse Faraday effect. This is a sure indicator of the need for an *O(3)* or other non-linear electrodynamics.

There is no known electric analogue of the inverse Faraday effect, suggesting that there is no $\boldsymbol{E}^{(3)}$ field as indicated by *O(3)* electrodynamics. The effect, as for its famous counterpart, the Faraday effect, is magnetic in nature and is mediated by the same Verdet constant [9].

The Stokes Theorem applied to $\boldsymbol{B}^{(3)}$ is clearly not to be found in the *U(1)* electrodynamics, and is the integral form of the $\boldsymbol{B}^{(3)}$ curl equation, $\nabla \times \boldsymbol{B}^{(3)} = \boldsymbol{0}$. This is simply a consequence of the fact that $\boldsymbol{B}^{(3)}$ is irrotational, and that the Stokes Theorem means that the curl of an irrotational vector field vanishes for any contour [8]. Again, such a result will not occur in *U(1)* electrodynamics because $\boldsymbol{B}^{(3)}$ is not defined there.

2.3.2.1 Self Consistency of Equations (1.2.57) and (1.2.50)

In order for the linearization scheme leading to Eqs. (1.2.59) and (1.2.60) to be applicable the solutions of Eq. (1.2.57) must be consistent with Eq. (1.2.50). In order to show this we write the covariant derivative as [5—8],

$$D_\mu = \partial_\mu - igM^a A_\mu^a , \tag{1.2.67}$$

so its action on the general m component field ψ_m is,

$$D_\mu \psi_m = \partial_\mu \psi_m - ig \left(M^a \right)_{mn} A_\mu^a \psi_n = \partial_\mu \psi_m - g \epsilon_{amn} A_\mu^a \psi_n . \tag{1.2.68}$$

Therefore the Feynman-Jacobi identity is

$$\partial_\mu \tilde{G}_m^{\mu\nu} - g\,\epsilon_{amn} A_\mu^a \tilde{G}_n^{\mu\nu} = 0 . \tag{1.2.69}$$

We know that the B cyclic theorem is constructed from,

$$\partial_\mu \tilde{G}_m^{\mu\nu} = 0 , \qquad m = (1), (2), (3) , \tag{1.2.70}$$

and that this implies Eq. (1.2.57), *i.e.*,

$$\epsilon_{amn} A_\mu^a \tilde{G}_n^{\mu\nu} = 0 . \tag{1.2.71}$$

At this point it is necessary to verify that Eq. (1.2.70) is self consistent with Eq. (1.2.71). For $m = (3)$, for example,

$$\partial_\nu \tilde{G}^{\mu\nu(3)*} - ig\,\epsilon_{(1)(2)(3)} A_\nu^{(1)} \tilde{G}^{\mu\nu(2)} = 0 , \tag{1.2.72}$$

with,

$$\tilde{G}^{03(3)*} = -\tilde{G}^{30(3)*} \neq 0 , \tag{1.2.73}$$

and with all other components zero. Therefore,

$$\partial_0 \tilde{G}^{03(3)*} - ig\left(A_0^{(1)} \tilde{G}^{03(2)} - A_0^{(2)} \tilde{G}^{03(1)} \right) = 0 . \tag{1.2.74}$$

This equation is consistent with,

$$\tilde{G}^{03(2)} = \tilde{G}^{03(1)} = 0 , \qquad A_0^{(1)} = A_0^{(2)} = 0 , \tag{1.2.75}$$

qed. For $m = 1$, Eq. (1.2.71) gives,

$$\epsilon_{(2)(1)(3)} A_\mu^{(2)} \tilde{G}_{(3)}^{\mu\nu} + \epsilon_{(3)(1)(2)} A_\mu^{(3)} \tilde{G}_{(2)}^{\mu\nu} = 0\,, \tag{1.2.76}$$

i.e.,

$$A_\mu^{(3)} \tilde{G}_{(2)}^{\mu\nu} = A_\mu^{(3)} \tilde{G}_{(1)}^{\mu\nu}\,. \tag{1.2.77}$$

For $m = (2)$,

$$A_\mu^{(1)} \tilde{G}_{(3)}^{\mu\nu} = A_\mu^{(3)} \tilde{G}_{(1)}^{\mu\nu}\,. \tag{1.2.78}$$

In vector notation, Eq. (1.2.78) gives,

$$\boldsymbol{A}^{(1)} \cdot \boldsymbol{B}^{(3)} = \boldsymbol{A}^{(3)} \cdot \boldsymbol{B}^{(1)}\,, \tag{1.2.79}$$

and

$$-A_0^{(1)} c\boldsymbol{B}^{(3)} + \boldsymbol{A}^{(1)} \times \boldsymbol{E}^{(3)} = -A_0^{(3)} c\boldsymbol{B}^{(1)} + \boldsymbol{A}^{(3)} \times \boldsymbol{E}^{(1)}\,. \tag{1.2.80}$$

Equation (1.2.79) is consistent with the fact that $\boldsymbol{A}^{(1)}$ is perpendicular to $\boldsymbol{B}^{(3)}$ and $\boldsymbol{A}^{(3)}$ to $\boldsymbol{B}^{(1)}$. Equation (1.2.80) simplifies to

$$\boldsymbol{A}^{(3)} \times \boldsymbol{E}^{(1)} = cA_0^{(3)} \boldsymbol{B}^{(2)*}\,, \tag{1.2.81}$$

which is consistent with the fact that the cross product of two polar vectors, $\boldsymbol{A}^{(3)}$ and $\boldsymbol{E}^{(1)}$, gives an axial vector $\boldsymbol{B}^{(2)*} = \boldsymbol{B}^{(1)}$ multiplied by $cA_0^{(3)}$, a scalar. To show that Eq. (1.2.81) is a component of the B cyclic theorem, and consistent with the fundamental definitions given in Chap. 1, use

$$\boldsymbol{A}^{(3)} = A_0^{(3)}\boldsymbol{k} = A_0^{(3)}\boldsymbol{e}^{(3)}, \tag{1.2.82}$$

so,

$$\boldsymbol{A}^{(3)} \times \boldsymbol{E}^{(1)} = A_0^{(3)}\left(\boldsymbol{k} \times \boldsymbol{E}^{(1)}\right) = cA_0^{(3)}\boldsymbol{B}^{(1)}, \tag{1.2.83}$$

a result which is consistent with

$$c\boldsymbol{B}^{(1)} = \boldsymbol{k} \times \boldsymbol{E}^{(1)}, \tag{1.2.84}$$

which in turn is a plane wave relation consistent with the fact that the B cyclic theorem is constructed from plane waves. This result is enough to show that Eq. (1.2.57) is consistent with Eq. (1.2.50). To reduce Eq. (1.2.83) to the B cyclic theorem use,

$$\boldsymbol{E}^{(1)} \rightarrow -ic\boldsymbol{B}^{(1)}, \qquad \boldsymbol{A}^{(3)} \rightarrow \frac{\boldsymbol{B}^{(3)}}{\kappa}. \tag{1.2.85}$$

The arrows are used in the above equations to denote that $\boldsymbol{E}^{(1)}$ is numerically the same as $-ic\boldsymbol{B}^{(1)}$ and that $\boldsymbol{A}^{(3)}$ is numerically the same as $\boldsymbol{B}^{(3)}/\kappa$. We do not use $\boldsymbol{E}^{(1)} = -ic\boldsymbol{B}^{(1)}$ because a polar vector cannot be equal to an axial vector in *O(3)*. Equations (1.2.85) are then examples of duality transformations, rather than equations. Equation (1.2.85a) occurs in the *U(1)* theory, *i.e.,* for transverse plane waves, Eq. (1.2.85b) occurs only in *O(3)* theory. Use of the transformations (1.2.85) in Eq. (1.2.83) produces

$$\boldsymbol{B}^{(3)} \times \boldsymbol{B}^{(1)} = iB^{(0)}\boldsymbol{B}^{(2)*}, \tag{1.2.86}$$

which is one component of the B cyclic theorem, *qed.* The overall result is that the B cyclic theorem linearizes Eq. (1.2.50) to Eq. (1.2.59) in a rigorously self consistent way.

2.3.2.2 Tensor to Vector Notation

Considering for example,

$$A_{\mu}^{(2)}\tilde{G}^{(1)\mu\nu} = A_{\mu}^{(1)}\tilde{G}^{(2)\mu\nu}, \tag{1.2.87}$$

and taking $\nu = 0$,

$$\begin{aligned} &A_0^{(2)}\tilde{G}^{(1)00} + A_1^{(2)}\tilde{G}^{(1)10} + A_2^{(2)}\tilde{G}^{(1)20} + A_3^{(2)}\tilde{G}^{(1)30} \\ &= A_0^{(1)}\tilde{G}^{(2)00} + A_1^{(1)}\tilde{G}^{(2)10} + + A_2^{(1)}\tilde{G}^{(2)20} + A_3^{(1)}\tilde{G}^{(2)30}, \end{aligned} \tag{1.2.88}$$

i.e.,

$$\boldsymbol{A}^{(2)} \cdot \boldsymbol{B}^{(1)} = \boldsymbol{A}^{(1)} \cdot \boldsymbol{B}^{(2)}, \tag{1.2.89}$$

in vector form..

Similarly, for $\nu = 1$,

$$\begin{aligned} &A_0^{(2)}\tilde{G}^{(1)01} + A_1^{(2)}\tilde{G}^{(1)11} + A_2^{(2)}\tilde{G}^{(1)21} + A_3^{(2)}\tilde{G}^{(1)31} \\ &= A_0^{(1)}\tilde{G}^{(2)01} + A_1^{(1)}\tilde{G}^{(2)11} + + A_2^{(1)}\tilde{G}^{(2)21} + A_3^{(1)}\tilde{G}^{(2)31}, \end{aligned} \tag{1.2.90}$$

i.e.,

$$-A_0^{(2)}B^{(1)1} - A_2^{(2)}E^{(1)3} + A_3^{(2)}E^{(1)2}$$

$$= -A_0^{(1)}B^{(2)1} - A_2^{(1)}E^{(2)3} + A_3^{(1)}E^{(2)2} \tag{1.2.91}$$

$$-A_0^{(2)}\boldsymbol{B}^{(1)} + \boldsymbol{A}^{(2)} \times \boldsymbol{E}^{(1)} = -A_0^{(1)}\boldsymbol{B}^{(2)} + \boldsymbol{A}^{(1)} \times \boldsymbol{E}^{(2)}$$

$$i.e.,\ \boldsymbol{A}^{(2)} \times \boldsymbol{E}^{(1)} = \boldsymbol{A}^{(1)} \times \boldsymbol{E}^{(2)} .$$

This result is consistent with the fact that

$$\boldsymbol{B}^{(2)} = \nabla \times \boldsymbol{A}^{(2)} , \tag{1.2.92}$$

qed.

Similarly self consistent results are found for $\nu = 2$ and for $\nu = 3$, demonstrating the rigorous self-consistency of *O(3)* electrodynamics in free space for the special case of plane waves $\boldsymbol{B}^{(1)} = \boldsymbol{B}^{(2)*}$ and for longitudinal phaseless $\boldsymbol{B}^{(3)}$.

2.3.3 Self Consistency of Equation (1.2.58)

Examination of the self consistency of Eq. (1.2.58) leads to the definition of a vacuum current that has no existence in *U(1)* theory. Linearization of Eq. (1.2.58) proceeds on the basis that in free space

$$\partial_\nu \boldsymbol{G}^{\mu\nu} = \boldsymbol{0} , \tag{1.2.93}$$

i.e.,

$$\partial_\nu G^{\mu\nu(1)} = \partial_\nu G^{\mu\nu(2)} = \partial_\nu G^{\mu\nu(3)} = 0 , \tag{1.2.94}$$

which is consistent with transverse plane waves; a phaseless longitudinal $\boldsymbol{B}^{(3)}$, and $\boldsymbol{E}^{(3)} = \mathbf{0}$ in free space.

This result implies that

$$g\boldsymbol{A}_\nu \times \boldsymbol{G}^{\mu\nu} = \frac{\boldsymbol{J}^\mu(\mathrm{vac})}{\epsilon_0}, \tag{1.2.95}$$

where $\boldsymbol{J}^\mu(\mathrm{vac})$ is a conserved current caused by the $\boldsymbol{B}^{(3)}$ field in free space and which does not exist in *U(1)* theory. We refer to it as the vacuum current, a polar vector in the *O(3)* symmetry internal gauge space whose scalar components in this space are polar four-vectors:

$$\left.\begin{aligned} J^{\mu(1)*} &= -ig\epsilon_0\left(A_\nu^{(2)}G^{\mu\nu(3)} - A_\nu^{(3)}G^{\mu\nu(2)}\right) \\ J^{\mu(2)*} &= -ig\epsilon_0\left(A_\nu^{(3)}G^{\mu\nu(1)} - A_\nu^{(1)}G^{\mu\nu(3)}\right) \\ J^{\mu(3)*} &= -ig\epsilon_0\left(A_\nu^{(1)}G^{\mu\nu(2)} - A_\nu^{(2)}G^{\mu\nu(1)}\right). \end{aligned}\right\} \tag{1.2.96}$$

Using Eqs. (1.1 .64) and (1.1.65) we can proceed to investigate the above cyclic relations for each μ.

For $\mu = 0$ we obtain, for example,

$$A_\nu^{(1)}G^{0\nu(3)} - A_\nu^{(3)}G^{0\nu(1)} = \frac{i}{g\epsilon_0}J^{0(2)*} = 0, \tag{1.2.97}$$

because all terms of

$$A_\nu^{(1)}G^{0\nu(3)} = A_\nu^{(3)}G^{0\nu(1)}, \tag{1.2.98}$$

are zero on both sides.

For $\mu = 1$,

$$A_\nu^{(1)} G^{1\nu(3)} - A_\nu^{(3)} G^{1\nu(1)} = \frac{i}{g\epsilon_0} J^{1(2)*} , \tag{1.2.99}$$

which reduces to

$$cA_Y^{(1)} B_Z^{(3)} = E_X^{(1)} A_0^{(3)} - cB_Y^{(1)} A_Z^{(3)} + \frac{i}{g\epsilon_0} J_X^{(1)} , \tag{1.2.100}$$

Using

$$E_X^{(1)} A_0^{(3)} = cB_Y^{(1)} A_Z^{(3)} , \tag{1.2.101}$$

we obtain

$$J_X^{(1)} = -icg\epsilon_0 A_Y^{(1)} B_Z^{(3)} = -i\epsilon_0 c \frac{A_Y^{(1)}}{A^{(0)}} B_Z^{(3)} , \tag{1.2.102}$$

which is a transverse current whose phase average is zero. It exists if and only if $\boldsymbol{B}^{(3)}$ is non-zero, and can be interpreted as a type of helicity current [8].

For $\mu = 2$ we obtain

$$A_\nu^{(1)} G^{2\nu(3)} - A_\nu^{(3)} G^{2\nu(1)} = \frac{i}{g\epsilon_0} J^{2(2)*} , \tag{1.2.103}$$

which reduces to

$$cA_1^{(1)} B^{3(3)} - A_0^{(3)} E^{2(1)} + cA_3^{(3)} B^{1(1)} = \frac{i}{g\epsilon_0} J^{2(1)} , \tag{1.2.104}$$

i.e.,

$$\frac{i}{g\epsilon_0}J^{2(1)} + A_0^{(3)}E^{2(1)} = cA_1^{(1)}B^{3(3)} + cA_3^{(3)}B^{1(1)} , \tag{1.2.105}$$

or

$$\frac{i}{g\epsilon_0}J_Y^{(1)} + A_0^{(3)}E_Y^{(1)} = -2cA_X^{(1)}B_Z^{(3)} , \tag{1.2.106}$$

and using

$$E_Y^{(1)} = -cB_X^{(1)} , \tag{1.2.107}$$

we obtain

$$J_Y^{(1)} = igc\epsilon_0 A_X^{(1)}B_Z^{(3)} . \tag{1.2.108}$$

Therefore Eq. (1.2.51) linearizes to Eq. (1.2.94) provided that there is a vacuum current which phase averages to zero. This can be identified as a helicity current due to $\boldsymbol{A}^{(1)}$ and $\boldsymbol{B}^{(3)}$. Equation (1.2.93) is a vector equation in the *O(3)* internal space and produces three scalar equations in this space, Eqs. (1.2.94). Those for indices (1) and (2) are the inhomogeneous Maxwell equations in free space and the third in vector form is Eq. (1.2.66), a result which shows that $\boldsymbol{B}^{(3)}$ is irrotational if $\boldsymbol{B}^{(1)}$ and $\boldsymbol{B}^{(2)}$ are plane waves or if $\boldsymbol{B}^{(1)}$ and $\boldsymbol{B}^{(2)}$ have imaginary phases opposite in sign. This is consistent with the B cyclic theorem, or vacuum magnetization.

To summarize, a theory of electrodynamics has been proposed based on the structure of general gauge field theory as used in particle and high energy physics. This theory linearizes self-consistently to the homogeneous and inhomogeneous Maxwell equations giving in the process a phase dependent vacuum current proportional directly to $\boldsymbol{B}^{(3)}$. This vacuum current is therefore zero in the *U(1)* theory: a self inconsistency of Maxwellian electrodynamics in the received view because if the electromagnetic field is assumed to propagate in vacuo at c, it must carry a

C negative influence at a finite velocity from one charge to another. This is a current which is always non-zero, yet which is set to zero in "charge free regions". In *O(3)* electrodynamics there is always a vacuum current which depend on the non-zero $\boldsymbol{B}^{(3)}$ component of the field. Therefore there are several ways in which the *O(3)* hypothesis is more self-consistent than the *U(1)* hypothesis. In other words, linearization as in Maxwellian electrodynamics removes a great deal of information and leads to self inconsistencies. The simplest possible type of non linear theory, based on the *O(3)* group symmetry, produces non-linear effects which are missing from the *U(1)* theory, and which are reinstated in that theory by hand. One of these is the vacuum current as demonstrated in this section; other examples include vacuum polarization and magnetization.

References

[1] R. Aldrovandi, in T.W. Barrett and D. M. Grimes, eds., *Advanced Electromagnetism, Foundations, Theory and Applications*, Chap. 1 (World Scientific, Singapore, 1995).

[2] J. D. Jackson, *Classical Electrodynamics* (Wiley, New York, 1962).

[3] W. K. H. Panofsky and M. Phillips, *Classical Electricity and Magnetism* (Addison-Wesley, Reading, 1962).

[4] L. H. Ryder, *Quantum Field Theory* (Cambridge University Press, Cambridge, 1987).

[5] M. W. Evans and J.-P. Vigier, *The Enigmatic Photon, Vol. 1: The Field* $\boldsymbol{B}^{(3)}$ (Kluwer Academic, Dordrecht, 1995).

[6] M. W. Evans and J.-P. Vigier, *The Enigmatic Photon, Vol. 2: Non-Abelian Electrodynamics* (Kluwer Academic, Dordrecht, 1995).

[7] M. W. Evans , J.-P. Vigier, S. Roy, and S. Jeffers, *The Enigmatic Photon, Vol. 3: Theory and Practice of the* $\boldsymbol{B}^{(3)}$ *Field* (Kluwer Academic, Dordrecht, 1996).

[8] M. W. Evans, J.-P. Vigier, and S.Roy, *The Enigmatic Photon, Vol. 4: New Directions* (Kluwer Academic, Dordrecht, 1998).

[9] M. W. Evans and S. Kielich eds., *Modern Nonlinear Optics*, Vols. 85(1), 85(2), 85(3) of Advances in Chemical Physics, I. Prigogine and S. A. Rice, eds. (Wiley Interscience, New York, 1993/1994/1997).

[10] K. Schwarzschild, *Göttinger Nachr., Math.-Phys. Klasse*, 1903, p.126

Chapter 3
Field-Matter Interaction

3.1 Introduction

In this chapter we demonstrate the fundamental $\boldsymbol{B}^{(3)}$ to one fermion interaction that leads to the phenomenon of radiatively induced electron spin resonance (*ESR*) and nuclear magnetic resonance (*NMR*). The Dirac equation was first solved to show the existence of radiatively induced fermion resonance (*RFR*) as reported in the third volume of this series [1]. The term responsible for the effect was isolated to be the novel interaction energy, the real valued and physical expectation value,

$$En = i\frac{e^2}{2m}\,\sigma\cdot\boldsymbol{A}\times\boldsymbol{A}^*\,, \tag{1.3.1}$$

where e/m is the charge to mass ratio of a fermion (electron or proton) and $i\boldsymbol{A}\times\boldsymbol{A}^*$ the real valued conjugate product of complex vector potentials in a circularly polarized electromagnetic field, considered to be classical in the manner first proposed by Dirac [2]. In Eq. (1.3.1), σ is the Z component of the Pauli matrix [3—6]. The interaction energy can be expressed in terms of the $\boldsymbol{B}^{(3)}$ field of the radiation as [1]

$$En = -\frac{e\hbar}{2m}\,\sigma\cdot\boldsymbol{B}^{(3)}\,, \tag{1.3.2}$$

and this is an *ESR* or *NMR* equation with the static magnetic field replaced by $\boldsymbol{B}^{(3)}$. All known *ESR* and *NMR* effects can therefore be induced by radiation rather than by a static magnetic field. The technique produces unprecedented resolving power because the resonance frequencies are proportional to I/ω^2 where I is the beam power density (or intensity in $W m^{-2}$) and ω beam angular frequency. With moderate microwave pumping, fermion resonance can be induced in theory in the visible, and picked up with an ordinary Fourier transform infra red-visible spectrometer acting as probe. This produces in theory a resolving power about one thousand to ten thousand times that available with magnet based *ESR* or *NMR* of any kind (including multi dimensional *ESR* and *NMR*) because the visible range is that much higher in frequency than the microwave (or gigahertz) range in which the current instruments operate.

This result indicates the existence and usefulness of the $\boldsymbol{B}^{(3)}$ field and is the fundamental spin-spin coupling between the photomagneton [7] (the photon's $\boldsymbol{B}^{(3)}$ field) and the fermion's half integral spin $\boldsymbol{B}^{(3)}$ proposed by Pauli [8] and Dirac [9]. Indications of the existence of $\boldsymbol{B}^{(3)}$ open the road to non-Abelian electrodynamics and non-local and superluminal interpretations [10] unknown in the traditional view [11].

In this chapter the above result is reproduced with several equations of motion, beginning with the Newton equation of a classical charged particle in a classical electromagnetic field; and ending with the quantum relativistic van der Waerden equation [12] for a two component spinor. The complete hierarchy of known equations of motion in physics produces the same *RFR* term, Eq. (1.3.1). It is a real, non-zero and physical ground state term in Rayleigh-Schrödinger perturbation theory [13]. The same type of coupling appears to have been recognized in principle by Pershan *et al.* [14] in 1966, during their establishment of the inverse Faraday effect, but these authors used higher order perturbation theory near optical resonance as did Li *et al.* [15] and others [16—20] in recent papers confirming the original proposal of *RFR* [21]. The key $\sigma \cdot \boldsymbol{A} \times \boldsymbol{A}^*$ coupling in higher order perturbation theory is clearly represented in Ref. [14] Eq. (8.6) and was confirmed by them empirically in paramagnetic, rare earth doped glass samples. These authors did not appear to realize however that the ground state term (1.3.1) is non-zero. This is the fundamental $\boldsymbol{B}^{(3)}$ term discussed in this chapter and occurs independently of any optical resonance, as in ordinary magnet based *ESR* and *NMR*. The great beauty of the new theory therefore is that one merely replaces $\boldsymbol{B}$ of the magnet by $\boldsymbol{B}^{(3)}$ of the electromagnetic field [1]. One can then proceed to understand the gallery of consequences as in the highly developed theory of *ESR* and *NMR*, but with a potential resolving power up to ten thousand times greater. In analogy, successful development would be the metaphorical equivalent of replacing the optical with the scanning tunneling electron microscope.

3.2 Classical Non-Relativistic Physics

In order to derive Eq. (1.3.1) in Newtonian physics, write the kinetic energy in *SU(2)* topology through the use of the Pauli matrix σ [8] and describe the field to particle interactions with the minimal prescription applied to a complex valued $\boldsymbol{A}$ representing the magnetic vector potential of the electromagnetic field [11]. Finally use ordinary complex algebra to extract the real valued and physically meaningful interaction kinetic energy corresponding to Eq. (1.3.1). The Newtonian kinetic energy of a classical charged particle interacting with the classical electromagnetic field in *SU(2)* topology is therefore the real part of

$$\begin{aligned} H_{\text{int}} &= \frac{1}{2m}\sigma\cdot(\boldsymbol{p}-e\boldsymbol{A})\sigma\cdot(\boldsymbol{p}-e\boldsymbol{A}^{*}) = \frac{1}{2m}\sigma\cdot\boldsymbol{p}\sigma\cdot\boldsymbol{p} \\ &\quad - \frac{e}{2m}(\sigma\cdot\boldsymbol{A}\,\sigma\cdot\boldsymbol{p}+\sigma\cdot\boldsymbol{p}\,\sigma\cdot\boldsymbol{A}^{*}) \\ &\quad + \frac{e^{2}}{2m}\sigma\cdot\boldsymbol{A}\sigma\cdot\boldsymbol{A}^{*}. \end{aligned} \tag{1.3.3}$$

Using the results,

$$\left.\begin{aligned} \sigma\cdot\boldsymbol{A}\sigma\cdot\boldsymbol{p} &= \boldsymbol{A}\cdot\boldsymbol{p} + i\sigma\cdot\boldsymbol{A}\times\boldsymbol{p}, \\ \sigma\cdot\boldsymbol{p}\sigma\cdot\boldsymbol{A}^{*} &= \boldsymbol{p}\cdot\boldsymbol{A}^{*} + i\sigma\cdot\boldsymbol{p}\times\boldsymbol{A}^{*}, \end{aligned}\right\} \tag{1.3.3a}$$

we can isolate the following terms from the right hand side of Eq. (1.3.3).

1) *Magnetic Dipole Term*

$$H_1 = -\frac{e}{2m}\boldsymbol{p}\cdot(\boldsymbol{A}+\boldsymbol{A}^*) \tag{1.3.3b}$$

$$= \frac{e}{2m}\boldsymbol{m}_0\cdot Re\,\boldsymbol{B}\,, \tag{1.3.3c}$$

where $\boldsymbol{m}_0$ is the magnetic dipole moment of the electron or proton and *Re* **B** is the real magnetic component of the electromagnetic field.

2) *Spin-Flip Term*

$$H_2 = -i\frac{e}{2m}\,\sigma\cdot\boldsymbol{p}\times(\boldsymbol{A}^* - \boldsymbol{A})\,, \tag{1.3.3d}$$

which for an electron or proton moving initially in the Z axis can be expressed as

$$H_2 = -e\frac{A^{(0)}}{\sqrt{2}}\,p_Z\,\sigma\cdot(\boldsymbol{j}\cos\phi + \boldsymbol{i}\sin\phi)\,, \tag{1.3.3e}$$

where

$$\phi = \omega t - \kappa Z = \omega\,(t - \frac{Z}{c})\,. \tag{1.3.3f}$$

If initially $\phi = 0$ the spin σ points in the Y axis; when $\phi = \pi/2$ it points in the X axis; when $\phi = \pi$ in the $-Y$ axis; when $\phi = 3\pi/2$ in the $-X$ axis and when $\phi = 2\pi$ back in the Y axis. So this confirms that H_2 is the spin-flip term used in all Fourier transform *ESR* and *NMR* instruments.

3) *Polarizability Term*

This is,

$$H_3 = \frac{e^2}{2m} \boldsymbol{A} \cdot \boldsymbol{A}^* = \frac{e^2}{2m} A^{(0)2} , \qquad (1.3.3g)$$

and is the basis of susceptibility theory [13].

4) *The RFR Term*

The *RFR* term, finally, is,

$$\left. \begin{aligned} H_4 &= i\frac{e^2}{2m} \sigma \cdot \boldsymbol{A} \times \boldsymbol{A}^* \\ &= -\frac{e^2}{2m} A^{(0)2} \sigma \cdot \boldsymbol{k} . \end{aligned} \right\} \qquad (1.3.3h)$$

All four terms have been observed empirically. Terms 1) to 3) are well known and term 4) was observed by Pershan et al. [14] in the paramagnetic inverse Faraday effect.

Thus Eq. (1.3.3) contains the spin-flip and *RFR* term in addition to the familiar and observable *O(3)* terms as found in a text such as that by Pike and Sarkav [22]. These terms rely for their existence on topology rather than quantum mechanics. It is well known [23] that *SU(2)* is homomorphic with *O(3)*, the usual rotation group of three dimensional space in Newtonian physics. However, the Clifford algebra underlying *SU(2)* gives more information, as advocated by Bearden *et al.* [24]. Our Newtonian result is consistent with the fact that Eq. (1.3.1) was obtained in the non-relativistic limit of the Dirac equation as a real expectation value [1].

3.3 Classical Relativistic Physics

It is a straightforward matter to repeat this simple exercise for classical relativistic physics because one can use the same minimal prescription in the Einstein equation written in *SU(2)* topology. For a free classical particle, the latter is

$$\gamma^{\mu} p_{\mu} \gamma^{\mu} p_{\mu} = m^2 c^2 , \tag{1.3.4}$$

where γ^{μ} is the Dirac matrix, p_{μ} the energy momentum four-vector, and c the speed of light in vacuo. The interaction of the classical electromagnetic field with the classical, relativistic, particle is described therefore by the equation of motion,

$$\gamma^{\mu} (p_{\mu} - eA_{\mu}) \gamma^{\mu} (p_{\mu} - eA_{\mu}^{*}) = m^2 c^2 , \tag{1.3.5}$$

which in Feynman's slash notation becomes [1]

$$(\not{p} - e\not{A})(\not{p} - e\not{A}^{*}) = m^2 c^2 . \tag{1.3.6}$$

The *RFR* term is [1] the real valued interaction energy,

$$En := \frac{e^2}{m} \not{A}\not{A}^{*} , \tag{1.3.7}$$

which includes term (1.3.1) of this chapter as part of a fully relativistic treatment,

$$\not{A}\not{A}^{*} = A_0 A_0^{*} - (\sigma \cdot \boldsymbol{A})(\sigma \cdot \boldsymbol{A}^{*}) = A_0 A_0^{*} - \boldsymbol{A} \cdot \boldsymbol{A}^{*} - i\sigma \cdot \boldsymbol{A} \times \boldsymbol{A}^{*} . \tag{1.3.8}$$

3.4 Non-Relativistic Quantum Physics

We can consider the Schrödinger Pauli equation [8],

$$\hat{H}\psi = En\,\psi\,, \tag{1.3.9}$$

in which the classical kinetic energy becomes an operator on a wavefunction which is a two component spinor in *SU(2)* topology. The usual operator replacements are used as follows:

$$\begin{aligned}
&p^{\mu} \rightarrow i\hbar\partial^{\mu}\,, && p_{\mu} \rightarrow i\hbar\partial_{\mu}\,,\\
&En \rightarrow i\hbar\frac{\partial}{\partial t}\,, && \boldsymbol{p} \rightarrow -i\hbar\nabla\,,\\
&p^{\mu} := \left(\frac{En}{c}, \boldsymbol{p}\right), && p_{\mu} := \left(\frac{En}{c}, -\boldsymbol{p}\right),\\
&\partial^{\mu} := \left(\frac{1}{c}\frac{\partial}{\partial t}, -\nabla\right), && \partial_{\mu} := \left(\frac{1}{c}\frac{\partial}{\partial t}, \nabla\right).
\end{aligned} \tag{1.3.10}$$

It is interesting to note that for a real valued $\boldsymbol{A}$ (static magnetic field problem of ordinary *ESR* and *NMR* [13]) the Schrödinger-Pauli equation produces the famous real expectation value,

$$En = -\frac{e\hbar}{2m}\sigma\cdot\boldsymbol{B}\,, \qquad \boldsymbol{B} = \nabla\times\boldsymbol{A}\,, \tag{1.3.11}$$

where $\hbar$ is the Dirac constant. This is the fundamental *ESR* or *NMR* term obtained in the non-relativistic quantum limit and has no classical equivalent because it depends for its existence on the operator rules (1.3.10). The Hamiltonian operator that produces result (1.3.11) is

$$\hat{H} = \frac{1}{2m}\sigma \cdot (\hat{\boldsymbol{p}} - e\boldsymbol{A})\sigma \cdot (\hat{\boldsymbol{p}} - e\boldsymbol{A}) + V, \qquad \hat{\boldsymbol{p}} = -i\hbar\hat{\nabla}, \tag{1.3.12}$$

where V is a potential energy.

In order to obtain the new *RFR* term (1.3.1) this operator becomes

$$\hat{H} = \frac{1}{2m}\sigma \cdot (\hat{\boldsymbol{p}} - e\boldsymbol{A})\, \sigma \cdot (\hat{\boldsymbol{p}} - e\boldsymbol{A}^{*}) + V, \tag{1.3.13}$$

and leads to the classical real valued term,

$$\hat{H}_{\mathrm{RFR}}\, \psi = i\frac{e^{2}}{2m}\, \sigma \cdot \boldsymbol{A} \times \boldsymbol{A}^{*}\, \psi, \tag{1.3.14}$$

which obviously has the same expectation value, Eq. (1.3.1). Therefore, unlike ordinary *ESR* and *NMR*, *RFR* depends on a term which *does* have a classical equivalent if we treat the field classically as did Dirac [2].

3.5 Relativistic Quantum Physics

The most straightforward route to relativistic quantum mechanics is through the replacement of p_{μ} in the Einstein equation (1.3.4) by its operator equivalent to give the van der Waerden equation of motion as detailed by Sakurai [8] for example,

$$\left(i\gamma^{\mu}\partial_{\mu}\right)\left(i\gamma^{\mu}\partial_{\mu}\right)\psi = \left(\frac{mc}{\hbar}\right)^{2}\psi. \tag{1.3.15}$$

Here ψ is a two component spinor and the equation is well known to be equivalent to the much better known Dirac equation involving a four component spinor. The *RFR* term emerges from the van der Waerden equation in the form,

$$\gamma^{\mu}\left(i\partial_{\mu}-eA_{\mu}\right)\gamma^{\mu}\left(i\partial_{\mu}-eA_{\mu}^{*}\right)\psi = \left(\frac{mc}{\hbar}\right)^{2}\psi. \tag{1.3.16}$$

The real and classical $e^2 \not{A}\not{A}^*$ is a simple multiplicative operator on the two component spinor, with the same, real, expectation value. This is also the case for the Dirac equation as given in Ref. 1, and in general for all *SU(2)* topology quantum mechanical equations.

3.6 Rayleigh-Schrödinger Perturbation Theory

In perturbation theory [13] the *RFR* term is a non-zero ground state term,

$$En = i\frac{e^2}{2m}\langle 0|\sigma\cdot \boldsymbol{A}\times\boldsymbol{A}^*|0\rangle + \text{second order terms.} \tag{1.3.17}$$

As shown recently by Li *et al.* [15] and by others [16—20] small second order *RFR* shifts also occur in second order corrections in perturbation theory, but term (1.3.17) is of far greater practical interest, because as shown in Ref. 1, it produces fermion resonances in the visible. Second order perturbation theory was also used by Pershan *et al.* [14] to produce the paramagnetic inverse Faraday effect, which they confirmed experimentally.

3.7 Discussion

In free space, the novel $\boldsymbol{B}^{(3)}$ field of *O(3)* symmetry electrodynamics is defined for one photon by,

$$\boldsymbol{B}^{(3)} := -i\frac{e}{\hbar}\boldsymbol{A}\times\boldsymbol{A}^*, \tag{1.3.18}$$

where e is the elementary charge [1]. Substituting this definition into Eq. (1.3.17) we find that the *RFR* term takes the same form precisely as the spin Zeeman effect produced by a static magnetic field,

$$En_{(\mathrm{RFR})} = -\frac{e\hbar}{2m}\langle 0|\sigma\cdot\boldsymbol{B}^{(3)}|0\rangle. \tag{1.3.19}$$

We need only replace $\boldsymbol{B}$ by $\boldsymbol{B}^{(3)}$ as defined in Eq. (1.3.18). Equation (1.3.19) is the fundamental spin-spin interaction between one photon and one fermion. For a free electron, the resonance frequency is straightforwardly calculated [1] from Eq. (1.3.19) to be

$$\omega_{\mathrm{res}} = \left(\frac{e^2\mu_0 c}{\hbar m}\right)\frac{I}{\omega^2} = 1.007\times10^{28}\frac{I}{\omega^2}, \tag{1.3.20}$$

where I is the pump beam power density in watts m^{-2} ($10{,}000$ watts m^{-2}) = 1.0 watt cm^{-2}), μ_0 the free space permeability in *SI* units. For the *H* atom, the Hamiltonian operator is well known to be,

$$\hat{H}_{(H\text{ atom})} = -\frac{\hbar^2}{2\mu}\nabla^2 + V, \tag{1.3.21}$$

where V denotes the classical Coulomb interaction between electron and proton and μ is the reduced mass,

$$\mu = \frac{m_e m_p}{m_e + m_p} \sim m_e, \tag{1.3.22}$$

where m_e and m_p are respectively the electron and proton masses. The resonance frequency in atomic H from Eq. (1.3.20) is therefore slightly shifted away from the free electron resonance frequency because the reduced mass is slightly different from the electron mass. The Hamiltonian operator

(1.3.22) for a monovalent alkali metal atom such as sodium (*Na*) must take account of the fact that there are several protons, neutrons and electrons arranged in orbitals according to the Pauli exclusion principle [13]. This atomic structure gives rise to the possibility of spin orbit coupling, spin-spin coupling between electrons, Fermi contact splitting, and hyperfine splitting as in *ESR* or *NMR* [13]. However, as a rule of thumb estimate, the outer or valence electron can be considered as superimposed on closed shells of inner electrons and a nucleus made up of protons and neutrons of a given reduced mass. To a first approximation, the Hamiltonian (1.3.22) can be used in which the sodium atom's reduced mass is slightly different from the free electron mass. This means that the main *RFR* resonance frequency in sodium is well estimated by Eq. (1.3.20) and so sodium vapor can be used in the experiment to detect *RFR*.

In order to detect *RFR* experimentally adjust conditions in the first instance so that,

$$\omega_{res} = \omega\,, \tag{1.3.23}$$

which is the auto-resonance condition in which the pump beam is absorbed at resonance because the pump frequency matches the resonance frequency precisely. Equation (1.3.20) simplifies to

$$\omega_{res}^3 = 1.007 \times 10^{28}\, I\,. \tag{1.3.24}$$

Therefore we can either tune ω_{res} for a given I or vice versa. Since auto-resonance must appear in the *GHz* if the pump frequency is in this range it is convenient to slightly modify the set up used by Deschamps *et al.* [25] in their detection of the inverse Faraday effect in plasma. They used a pulsed microwave signal at $3.0\,GHz$ from a klystron delivering megawatts of power over 12 microseconds with a repetition rate of 10 *Hz*. The TE_{11} Mode was circularly polarized with a polarizer placed inside a circular waveguide of $7.5\,cm$ diameter. The plasma sample was created by a very intense microwave pulse and held in a pyrex tube inserted coaxially in the waveguide of 6.5 *cm* diameter and length 20.0 *cm*. The section of the waveguide surrounding the tube was made of nylon internally coated with a 20 *micron* layer of copper. The inverse Faraday effect was then picked up with Faraday induction [25].

To detect *RFR* change the sample to sodium vapor, which is easily prepared and held in the sample tube. Equation (1.3.24) predicts that resonance occurs at 3.0 *GHz* if I is tuned to 0.0665 *watts* cm^{-2}. For a circular waveguide of 7.5 *cm* diameter this requires only 2.94 *watts* of *CW* power from the klystron at 3.0 *GHz*. In deriving Eq. (1.3.20) it has been assumed that [1,26]

$$I = \frac{c}{\mu_0} B^{(0)2} . \tag{1.3.25}$$

This is a simple theoretical estimate and it is strongly advisable that I can be tuned over a considerable range around 2.94 *watts* to allow for unforeseen discrepancies between Eq. (1.3.25) and the actual experimental beam intensity generated by the apparatus. Once the main resonance is detected however, further refinements can follow, making full use of contemporary electronics. To repeat the experiment with atomic H or with the free electron gas is likely to be more difficult purely because of sample handling problems. The experiment should be repeated after auto-resonanc is detected to demonstrate the major advantage of *RFR* by pulsing the pump beam for increased power density at the same frequency and by using Eq. (1.3.24) to estimate the resonance frequency. A sample of expected results is given in Table 3.1. As can be seen it is possible in theory to produce *ESR* (and *NMR*) in the visible range, with a four-order of magnitude increase in resolving power over current magnet based techniques.

Table 1. *RFR* Frequencies for a 3.0 GHz Pump for given *I*

Pump Intensity I (*watts* cm^{-2})	Resonance Frequency
10.0	15.04 cm^{-1} (Far infra red)
100.0	150.4 cm^{-1} (Far infra red)
1,000.0	1,504 cm^{-1} (Infra red)
10,000.0	15,040 cm^{-1} (Visible)
100,000.0	150,040 cm^{-1} (Ultra violet-X ray)

For a 3.0 *GHz* circularly polarized pump pulse of 10 *kwatts* cm^{-2} the *RFR* frequency is at 15,040 cm^{-1} in the visible, and can be detected with a Fourier transform infra red-visible spectrometer such as a fully computerized Bruker *IFS 113v*. The detector of the spectrometer must be fast enough to record an interferogram during the microsecond interval of the microwave pulse. Therefore pulse repetition and computer based refinement is necessary for good quality data. The pump should be kept as homogeneous and noise free as possible, but because of the I/ω^2 dependence of *RFR* , simple Maxwell-Boltzmann theory [1] shows that conditions can be adjusted to produce a much larger population difference between up and down fermion spins than in magnet based *ESR* or *NMR*. Therefore this alleviates the well known problem of magnet homogeneity in magnet based *ESR* and *NMR*, a problem which is due to a small (one part in a million) population difference. In *RFR* the latter can easily exceed 20% [1] at a conservative estimate for moderate pump power of ten *watts* order of magnitude. The complete *ESR* spectrum of sodium vapor can therefore be taken, in theory, in the infra red or visible. This is terra incognita in magnet based technology, which is reaching its design limit. The whole process can then be repeated for *NMR* and *MRI*.

The characteristic and key I/ω^2 coefficient of our theory [1] appears also in the second order perturbation theory of Harris and Tinoco [17], their p. 9291, second column, premultiplied by a factor. These authors miss the first order or ground state term (1) and in consequence their theory falls short of empirical indications by Warren *et al.* [27] by eight orders of magnitude. Straightforward estimates [1] based on Eq. (1.3.1) applied to *NMR* fall in the order of magnitude of the data obtained by Warren *et al.* [27] by visible frequency irradiation of molecular liquids with various circularly polarized lasers, including an argon ion laser at 528.7 *nm*, 488 *nm*, and 476.5 *nm*. Accounting simply for the different g factors of the proton and electron, Eq. (1.3.1) applied to *NMR* [1] produces very tiny shifts of 0.12, 0.10 and 0.098 *Hz* respectively for the three argon ion laser frequencies quoted above and for an intensity of ten *watts* per square *centimeter*, approaching the highest *CW* intensities used by Warren *et al.* [27] in important and pioneering experiments at Princeton following our early theory [21,28] which also missed the key term (1.3.1) introduced finally in Ref. 1. Equation (1.3.1) now shows now why Warren *et al.* [27] were not able to obtain more than indications of *RFR* shifts, both in proton and ^{13}C Fourier transform and two dimensional *NMR*. In ^{13}C *NMR* the mass of the ^{13}C nucleus is an order of magnitude heavier than in ^{1}H *NMR* and the shifts from Eq. (1.3.1), all other factors being equal, are in consequence an order of magnitude smaller, in the 0.01 *Hz* range — too small to be detected, as found experimentally [27]. The remedy is also given by Eq. (1.3.1), which is to replace the lasers with pulsed or *CW* microwave generators for about the same I. Their effort [27] nevertheless remains as a landmark in the field.

Finally, Li *et al.* [15] have shown that even in second order perturbation theory of the type used by Harris and Tinoco [16,17], or Buckingham *et al.* [18,19], large *RFR* shifts of up to 10 *MHz* are possible using pump lasers tuned near to optical resonance. Systematic development of *RFR*, first proposed by the present author in Ref. 21 and in several consequent papers [28], is clearly going to be highly beneficial to chemical physics and medicine unless all the equations of physics are misleading or unless some unforeseen technical difficulty occurs. With contemporary technology it is unlikely that such a difficulty, if it occurred, could not be overcome. Philosophically the whole process can be thought of as stemming from the $\boldsymbol{B}^{(3)}$ (Evans-Vigier) field of *O(3)* electrodynamics [1], which for one photon, is the fundamental photomatic [29].

References

[1] M. W. Evans, J.-P. Vigier, S. Roy, and S. Jeffers, *The Enigmatic Photon, Volume Three: Theory and Practice of the* $\boldsymbol{B}^{(3)}$ *Field,* Chaps. 1 and 2,. (Kluwer Academic, Dordrecht, 1996).

[2] P. A. M. Dirac, *Quantum Mechanics,* 4th edn. revised, (Oxford University Press, Oxford, 1974).

[3] L. H. Ryder, *Quantum Field Theory,* 2nd edn., (Cambridge University Press, Cambridge,1987).

[4] J. D. Bjorken and S. D. Drell, *Relativistic Quantum Mechanics* (McGraw Hill, New York, 1964).

[5] C. Itzykson and J. B. Zuber, *Quantum Field Theory* (McGraw Hill, New York, 1980).

[6] R. R. Ernst, G. Bodenhausen, and A. Wokaun, *Principles of Nuclear Magnetic Resonance in One and Two Dimensions* (Oxford University Press, Oxford, 1987).

[7] M. W. Evans, *Physica B* **182**, 227, 237 (1992); **183**, 103 (1993); **190**, 310 (1993); *Physica A* **215**, 605 (1995).

[8] J. J. Sakurai, *Advanced Quantum Mechanics* 11th printing, (Addison Wesley, New York, 1967) Chap. 3.

[9] V. B. Berestetskii, E. M. Lifshitz, and L. P. Pitaevski, *Relativistic Quantum Theory* (Pergamon, Oxford, 1971).

[10] M. W. Evans, J.-P. Vigier, and M. Meszaros, eds., *The Enigmatic Photon, Volume Five: Part 2, Selected papers of Erasmo Recami,* (Kluwer Academic, Dordrecht, *in preparation*).

[11] J. D. Jackson, *Classical Electrodynamics,* 1st edn. (Wiley, New York, 1962).

[12] B. L. van der Waerden, *Group Theory and Quantum Mechanics* (Springer Verlag, Berlin, 1974).

[13] P. W. Atkins, *Molecular Quantum Mechanics,* 2nd edn., (Oxford University Press, Oxford, 1983), pp. 383 ff.

[14] P. S. Pershan, J. van der Ziel, and L. D. Malmstrom, *Phys. Rev.* **143**, 574 (1966).

[15] L. Li, T. He, X. Wang, and F.-C. Liu, *Chem. Phys. Lett.* **268**, 549 (1997).
[16] R. A. Harris and I. Tinoco, *Science* **259**, 835 (1993).
[17] R. A. Harris and I. Tinoco, *J. Chem. Phys.* **101**, 9289 (1994).
[18] A. D. Buckingham and L. C. Parlett, *Chem. Phys. Lett.* **243**, 15 (1995).
[19] A. D. Buckingham and L. C. Parlett, *Science* **264**, 1748 (1994).
[20] M. W. Evans, *Found. Phys. Lett.* **8**, 563 (1995).
[21] M. W. Evans, *J. Mol. Spect.* **143**, 327 (1990); *Chem. Phys.* **157**, 1 (1991); *J. Phys. Chem.* **95**, 2256 (1991); *J. Mol. Liq.* **49**, 77 (1991); *Int. J. Mod. Phys. B* **5**, 1963, invited rev., (1991); *J. Mol. Spect.* **146**, 143 (1991); *Physica B* **168**, 9 (1991); *Adv. Chem. Phys.* **51**, 361—702 , invited rev. (1992); M. W. Evans, S. Woźniak, and G. Wagnière, *Physica B* **173**, 357 (1991); **175**, 416 (1991).
[22] E. R. Pike and S. Sarkav, *The Quantum Theory of Radiation*, Example 3.4, (Oxford University Press, Oxford, 1995).
[23] T. W. Barrett and D. M. Grimes, eds., *Advanced Electromagnetism, Foundations, Theory and Applications* (World Scientific, Singapore, 1995).
[24] T. W. Bearden, www.europa.com/~ rsc/physics, 1997 to present.
[25] J. Deschamps, M. Fitaire, and M. Lagoutte, *Phys. Rev. Lett.* **25**, 1330 (1970).
[26] L. D. Barron, *Molecular Light Scattering and Optical Activity*, Eq. (2.2.16) (Cambridge University Press, Cambridge, 1982).
[27] W. S. Warren, S. Mayr, D. Goswami, and A. P. West., Jr., *Science* **255**, 1683 (1992); **259**, 836 (1993).
[28] M. W. Evans and S. Kielich, eds., *Modern Nonlinear Optics*, Vol. 85(2) of *Advances in Chemical Physics,* I. Prigogine and S. A. Rice, eds., pp. 51 ff (Wiley Interscience, New York, 1992, 1993, and 1997 (paperback printing)).; also M. W. Evans, *Physica B* **182**, 118 (1992); **176**, 254 (1992); **179**, 157 (1992); S. Wozniak, M. W. Evans, and G. Wagniere, *Mol. Phys.* **75**, 81, 99 (1992).
[29] M. W. Evans and A. A. Hasanein, *The Photomagneton in Quantum Field Theory* (World Scientific, Singapore, 1994); M. W. Evans and

J.-P. Vigier, *Classical Electrodynamics and the* $\boldsymbol{B}^{(3)}$ *Field* (World Scientific, *in preparation*).

Part II

Historical Development of O(3) Electrodynamics: Selected Papers

Paper 1

Ultra Relativistic Inverse Faraday Effect

In the ultra relativistic limit of the inverse Faraday effect it is shown that the magnetization of the sample by the circularly polarized electromagnetic field becomes directly proportional to the $\boldsymbol{B}^{(3)}$ field of the radiation. Observation of such an effect is direct observation of the $\boldsymbol{B}^{(3)}$ field.

Key words: Inverse Faraday effect; Ultra relativistic limit; $\boldsymbol{B}^{(3)}$ field.

1.1 Introduction

The simple geometrical hypothesis of a $\boldsymbol{B}^{(3)}$ field of electromagnetic radiation is supported by the inverse Faraday effect, which to date has been verified experimentally in the non-relativistic limit, defined by relatively low intensity and relatively high frequency [1-5]. In such a limit the observable magnetization is proportional to beam intensity through the factor $B^{(0)}\boldsymbol{B}^{(3)}$, where $\boldsymbol{B}^{(3)} = B^{(0)}\boldsymbol{k}$. The effect is conventionally interpreted through the conjugate product of plane waves in the vacuum, within the traditional framework of Maxwell-Lorentz theory. The *U(1)* constraint imposed on such a theory implies that there is no $\boldsymbol{B}^{(3)}$ field in the vacuum [6—10]. However, if this group constraint is removed, the B cyclic theorem shows that the conjugate product is proportional to the recently inferred $\boldsymbol{B}^{(3)}$ field by ordinary three dimensional geometry [11]. Such group constraints reduce the generality even of the restricted, linear Maxwell-Lorentz-Cartan electrodynamics, and are, in the last analysis, subjective [12—15]. There is no reason therefore to reject the existence of the $\boldsymbol{B}^{(3)}$ field on the basis of the *U(1)* group restriction applied to a linear theory of electrodynamics. By hypothesis, $\boldsymbol{B}^{(3)}$ is a field of non-linear electrodynamics, and is not hypothesized in Maxwell-Lorentz electrodynamics.

Furthermore, the B Cyclic theorem is by its very nature a non-Maxwellian construct, which shows geometrically that the curl of the $\boldsymbol{B}^{(3)}$ field is zero in the vacuum. This prediction has recently been confirmed experimentally [16]. The theorem has been shown to be rigorously Lorentz covariant [17], and therefore quantizes to a $\hat{C}\hat{P}\hat{T}$ conserving field theory. It therefore has merit in special relativity and quantum mechanics. On these criteria, the $\boldsymbol{B}^{(3)}$ field is as valid as any other field component in theories that are Lorentz covariant. There is therefore no reason to assert that $\boldsymbol{B}^{(3)} = \mathbf{0}$, and no reason to assert that it is a field of Maxwell-Lorentz electrodynamics. If observed experimentally therefore, it signifies an advance in the basic structure of electrodynamic theory. The B Cyclic theorem is more fundamental in nature than the Maxwell-Lorentz-Cartan theory, because the theorem is, tautologically, an angular momentum operator relation, i.e., within $\hbar$, a relation between rotation generators of space itself [18]. It is therefore as fundamental as a geometrical hypothesis such as the Pythagorean theorem. For the first time, it applies relativistically correct, Lorentz covariant, geometry to *three* magnetic field (or rotation generator) components in vacuo interlinked by the structure of space-time. If we break this link, we automatically impose a subjective constraint, and change the ordinary topology of space-time. There is no way of arguing against the $\boldsymbol{B}^{(3)}$ field using a *model* of electrodynamics, especially a linear model, because the latter is inevitably constructed in the same space-time. The B Cyclic theorem is as fundamental as the Noether theorem. Space-time geometry, and the concomitant relation between fields, is valid irrespective of any model of electrodynamics, such as that of Maxwell. The B Cyclic theorem is a geometrical relation between three magnetic field components, all three of which are propagating at c through the vacuum. Such a concept obviously does not exist in Maxwell-Lorentz electrodynamics, yet is Lorentz covariant [17] and $\hat{C}\hat{P}\hat{T}$ conserving. Any criticism of $\boldsymbol{B}^{(3)}$ based on Maxwell-Lorentz electrodynamics is therefore misplaced from the beginning [19]. Such criticism applies an inadequate linear model to a fundamental, topological, non-linear and very fundamental new theorem.

In this paper it is shown that in the ultra relativistic limit of the inverse Faraday effect, the magnetization observable in the sample is proportional directly to $\boldsymbol{B}^{(3)}$, to no other magnetic field component, and is a direct demonstration of its existence. Due to the relation between $\boldsymbol{B}^{(3)}$ and the transverse plane waves [20] this result can be obtained from the relativistic Hamilton-Jacobi equation [21] in a limit of very low frequency and very high intensity. Furthermore, this limit is experimentally accessible [22]. In Sec. 1.2, the reasoning leading to this result is reviewed in terms of delayed action at a distance theory, which was shown by Schwarzschild [23] to be fully equivalent to Maxwell-Lorentz theory. This picture is not adequate for the interpretation of the inverse Faraday effect (*IFE*) however, because as we have seen, $\boldsymbol{B}^{(3)}$ is not defined in Maxwell-Lorentz theory, but is a very useful way of thinking of the inverse Faraday effect reduced to its essence. To properly define $\boldsymbol{B}^{(3)}$, a non-linear theory of electrodynamics is the minimum requirement. In Sec. 1.3, the ultra-relativistic limit is developed for one electron, and the equation given showing the direct relation between $\boldsymbol{B}^{(3)}$ and magnetization. The latter is longitudinally directed and can be proportional only to a longitudinally directed magnetic flux density propagating through a vacuum. This is the $\boldsymbol{B}^{(3)}$ field of the radiation.

1.2 Delayed Action at a Distance

As pointed out by Ritz [23], in an elegant criticism of Maxwell-Lorentz electrodynamics, the latter is exactly equivalent to delayed elementary action at a distance. The inverse Faraday effect can be understood in terms of elementary actions without the use of intervening fields. In the simplest case a circling electron in a transmitter radiates into the vacuum and a time t later an electron in a receiver is set into circular motion. Magnetization in the transmitter becomes magnetization in the receiver, both vectors being longitudinally directed in the Z axis. To describe this process mathematically requires the use of the elementary actions in a relativistic equation of motion, the most convenient one for this purpose is the relativistic Hamilton-Jacobi equation, as pointed out by Landau and Lifshitz [24]. The electronic motion set up in the receiver by the field is circular motion, so the elementary actions must be introduced in such a way as to reproduce this experimental fact, which can be inferred from the observation of magnetization in the Z axis of the receiver due to the electron circling about the Z axis.

When the calculation is carried out [25], the final result can be expressed in terms of a magnetic field $\boldsymbol{B}^{(3)}$, but it can also be expressed as an angular momentum due to delayed elementary action at a distance. In the last analysis it is simply a transfer of angular momentum from the transmitter to the receiver. The intervening agent is postulated to be a field, whose mathematical structure is determined by the partial differential equations known as Maxwell's equations. However, as shown by Ritz [23], these equations are no more than model relations between space-time components, whereas the B Cyclic theorem depends on no model.

The ultra relativistic limit being considered here is one in which the observed magnetization is directly proportional to $\boldsymbol{B}^{(3)}$ in the field theory. This is a simple result obtained after a long and complicated calculation based on the use of elementary action in the Hamilton-Jacobi equation. The same calculation produces the well known non - relativistic limit, which has been confirmed experimentally [1—5]. The elementary action is introduced in such a way as to spin the electron in the receiver, and in such a way as to reproduce the time it takes for the signal to reach the receiver from the transmitter. A combination of spinning and translating motions, combined with a time delay, means that there is a phase present, which is the electromagnetic phase. This appears only in the transverse components of the elementary action, which when put into the Hamilton-Jacobi equation of the electron in the receiver, produces the required circling motion. Because of the B Cyclic Theorem, this is identical with spinning the electron with a $\boldsymbol{B}^{(3)}$ field, and this is exactly what the result gives us in the ultra relativistic limit. The magnetization is directly proportional to $\boldsymbol{B}^{(3)}$. It is simply a magnetic field strength in the receiver produced by the magnetic flux density $\boldsymbol{B}^{(3)}$ of the vacuum, produced in turn by a magnetic field strength in the transmitter.

1.3 Calculation from the Hamilton-Jacobi Equation

The calculation given in Ref. 25 for the relativistic inverse Faraday effect is modified here for the ultra-relativistic limit by correcting the gyromagnetic ratio by the relativistic factor γ . In Gaussian units, the necessary expression is given by Talin *et al.* [26] in their Eq. (3),

$$M_Z = -\frac{|e|}{2mc\gamma} L_Z . \tag{2.1.1}$$

This corrects the usual gyromagnetic ratio [26], $e/2mc$ in Gaussian units. The origin of this correction is given by Talin *et al.* [26] in their Eq. (12),

$$M_Z = \frac{1}{V}\int \frac{dr}{2c} (\boldsymbol{r} \times \boldsymbol{j}(\boldsymbol{r},t))_Z , \tag{2.1.2}$$

where M_Z is magnetization, V the sample volume, $\boldsymbol{r}$ the relativistic radius of gyration and $\boldsymbol{j}(\boldsymbol{r},t)$ is a symmetrized current density operator. From this equation they develop their gauge invariant expression (38), which is proportional to the conjugate product of vector potentials, $\boldsymbol{A} \times \boldsymbol{A}^*$. This is the B Cyclic theorem in plasma,

$$\boldsymbol{M}^{(3)*} = -ig' \boldsymbol{A}^{(1)} \times \boldsymbol{A}^{(2)}, \tag{2.1.3}$$

where g' is relativistic and depends on the plasma properties.

In Ref. 25, the relativistic Hamilton-Jacobi equation is used to calculate the angular momentum set up in one electron by a circularly polarized, classical, electromagnetic field. The result is the same as that of Talin *et al.* , except that *SI* units are used in Ref. 25. However, in Ref. 25, the non-relativistic gyromagnetic ratio was used to relate this angular momentum to the magnetization. In *SI* units this is $-e/2m$ as given in Eq. (9.3.1) of Atkins [27], and as used to define the usual Bohr magneton $e\hbar/2m$. From Eq. (1.3) of Talin *et al.* [26] it can be seen that the gyromagnetic ratio itself needs to be corrected relativistically under some conditions. Therefore the Bohr magneton is not a constant, it depends on the relativistic factor γ as defined by Talin *et al.* [26].

When the necessary correction is made to the Bohr magneton, the SI magnetization becomes

$$\boldsymbol{M}^{(3)} = -\frac{e^3 c^2}{2m^2\omega^3 V}\left(\frac{1}{1+\left(\frac{eB^{(0)}}{m\omega}\right)^2}\right) B^{(0)} \boldsymbol{B}^{(3)}. \tag{2.1.4}$$

This is the magnetization in amps per meter (coulombs per second per meter) caused in one electron by a circularly polarized electromagnetic field. The magnitude of the magnetic flux density of the field is $B^{(0)}$, and its angular frequency is ω. The charge to mass ratio of the electron is e/m. The sample volume is V in cubic meters. Finally $\boldsymbol{B}^{(3)} := B^{(0)}\boldsymbol{k}$ where $\boldsymbol{k} = \boldsymbol{e}^{(3)}$. Equation (2.1.4) is valid over the whole range of existence of the inverse Faraday effect, from non-relativistic to ultra-relativistic. In the non-relativistic limit we obtain the same result as Ref. 25,

$$\boldsymbol{M}^{(3)} \xrightarrow{eB^{(0)} \ll m\omega} -\frac{1}{V}\left(\frac{e^3c^2}{2m^2\omega^3}\right)B^{(0)}\boldsymbol{B}^{(3)} , \tag{2.1.5}$$

and the magnetization is proportional to intensity. This is the original inverse Faraday effect, first observed in 1965 [1] and repeated several times [4—5].

In the ultra-relativistic limit, $eB^{(0)} \gg m\omega$, Eq. (2.1.4) becomes,

$$\boldsymbol{M}^{(3)} \xrightarrow{eB^{(0)} \gg m\omega} -\frac{1}{V}\left(\frac{ec^2}{2\omega}\right)\boldsymbol{e}^{(3)} . \tag{2.1.6}$$

This seems to be independent of $\boldsymbol{B}^{(3)}$, but recall [27] that the magnetic dipole moment of the field is,

$$|\boldsymbol{m}^{(3)}| = \frac{ec^2}{\omega} = \frac{V_0}{\mu_0}|\boldsymbol{B}^{(3)}| . \tag{2.1.7}$$

Therefore in the ultra-relativistic limit,

$$\boldsymbol{M}^{(3)} = -\frac{1}{2}\frac{1}{\mu_0}\left(\frac{V_0}{V}\right)\boldsymbol{B}^{(3)} , \tag{2.1.8}$$

where V_0 is the local volume element used to describe electromagnetic energy in vacuo, and μ_0 is the vacuum permeability.

Equation (2.1.8) is clear proof that in the ultra-relativistic limit, the magnetization is directly proportional to the $\boldsymbol{B}^{(3)}$ field of the radiation. If $\boldsymbol{B}^{(3)}$ were zero there would be no magnetization, and this is inconceivable, because the structure of Eq. (2.1.4) is valid over the entire range from non-relativistic to ultra-relativistic. This means that there must be an ultra-relativistic effect because there is an observable non-relativistic effect. Equation (2.1.4), except for units, is identical with the second part of Eq. (3) of Talin *et al.* [26].

Under all conditions,

$$\nabla \times \boldsymbol{M}^{(3)} = \nabla \times \boldsymbol{B}^{(3)} = \boldsymbol{0}\,. \tag{2.1.9}$$

Equation (2.1.6) or (2.1.8) is potentially very useful for applications as discussed below.

1.4 Discussion

The ultra-relativistic limit (2.1.6) can be reached experimentally [22] in the laboratory with standard apparatus. It is expected that the effect can be observed in all materials with power line apparatus as standard in the industry. From Eq. (2.1.6), which is more suitable for electrical engineering applications than Eq. (2.1.8), it is clear that the expected effect is inversely proportional to the angular frequency and the sample volume. Therefore in order to maximize the effect it is necessary to maximize the power density of the radiation field, minimize its frequency and minimize the sample volume. Although the power density does not seem to be present in Eq. (2.1.6), recall that it is the limit of Eq. (2.1.4). The power density reveals itself through the product $V^{(0)}\boldsymbol{B}^{(3)}$ in Eq. (2.1.8). It is easily checked that both Eqs. (2.1.6) and (2.1.8) have the required units of coulombs per second per meter.

These are the *SI* units both of magnetization and of magnetic field strength. So in the ultra relativistic limit the entire magnetic dipole moment of the field is transferred to the electron. As shown by Eq. (2.1.6), there is no intensity dependence in this limit. Therefore a constant magnetization is set up in a sample, such as a ferrite core of a power line design [22] and this phenomenon is entirely new and unexplored, probably with many applications. Although $\boldsymbol{B}^{(3)}$ appears in Eq. (2.1.8), it is multiplied by $V^{(0)}$, and the product $V^{(0)}\boldsymbol{B}^{(3)}$ has no power density dependence. The latter enters indirectly however because we are considering a extreme high power density-low frequency limit of Eq. (2.1.4). Remarkably, this limit is easily accessible in the laboratory and in applications [22].

The conclusion is that if $\boldsymbol{B}^{(3)}$ were zero, there would be no observable magnetization and this contradicts experience in the non-relativistic limit of the main equation (2.1.4). Therefore there can be no further doubt that $\boldsymbol{B}^{(3)}$ is non-zero empirically. It is a non- Maxwellian field.

Acknowledgments

The Alpha Foundation is thanked for an honorary praesidium membership and professorship. Prof. David Cykanski of Worcester Polytechnic Institute is thanked for discussions and a description of the experimental method in Ref. 22.

References

[1] J. P. van der Ziel, P. S. Pershan, and L. D. Malmstrom, *Phys. Rev. Lett.* **15**, 190 (1965); *Phys. Rev.* ***143**, 574 (1966).*

[2] J. Deschamps, M. Fitaire, and M. Lagoutte, Phys. Rev. Lett. **25**, 1330 (1970); *Rev. Appl. Phys.* 7, 155 (1972).

[3] T. W. Barrett, H. Wohltjen, and A. Snow, *Nature* **301**, 694 (1983).

[4] B. A. Zon, V. Yu. Kuperschmit, G. V. Pakhomov, and T. T. Urazbaev, *JETP Lett.* **45**, 272 (1987).

[5] G. H. Wagnière, *Linear and Non-Linear Optical Properties of Molecules*, (VCH, Basel, 1993).

[6] M. W. Evans, *Physica B* **182**, 227, 237 (1992).

[7] M. W. Evans and S. Kielich, eds., *Modern Nonlinear Optics*, Vols. 85(1—3) of *Advances in Chemical Physics*, I. Prigogine and S. A. Rice, eds. (Wiley Interscience,, New York, 1992, 1993, 1997 (paperback)).
[8] M. W. Evans and A. A. Hasanein, *The Photomagneton in Quantum Field* Theory (World Scientific, Singapore, 1994).
[9] M. W. Evans and J.-P. Vigier, *The Enigmatic Photon, Vol. 1: The Field* $\boldsymbol{B}^{(3)}$ (Kluwer Academic, Dordrecht, 1994).
[10] M. W. Evans and J.-P. Vigier, *The Enigmatic Photon, Vol. 2: Non-Abelian Electrodynamics* (Kluwer Academic, Dordrecht, 1995).
[11] M. W. Evans, J.-P. Vigier, and, S. Roy, eds., *The Enigmatic Photon, Vol. 4: New Directions* (Kluwer Academic, Dordrecht, 1998).
[12] R. M. Kiehn, G. P. Kiehn and R. B. Roberds, *Phys. Rev. A* **43**, 5665 (1991).
[13] R. M. Kiehn, *Int. J. Eng. Sci.* **14**, 749 (1976).
[14] R. M. Kiehn, *J. Math. Phys.* **18**, 614 (1977).
[15] R. M. Kiehn, *Phys. Fluids* **12**, 1971 (1969).
[16] M. Y. A. Raja, W. N. Sisk, M. Yousaf and D. Allen, *Appl. Phys. B* **64**, 79 (1997).
[17] V. V. Dvoeglazov, *Found. Phys. Lett.* **10**, 383 (1997).
[18] L. H. Ryder, *Quantum Field Theory* (Cambridge University Press, Cambridge, 1987).
[19] E. Comay, *Chem. Phys. Lett.* **261**, 601 (1996); M. W. Evans and S. Jeffers, *Found. Phys. Lett.* **9**, 587 (1996) (reply).
[20] M. W. Evans, *Physica B* **190**, 310 (1993).
[21] Ref. 9, Chap. 12.
[22] D. Cyganski, School of Engineering, Worcester Polytechnic Institute, Worcester, Mass., USA, e mail communications, Sept./Oct., 1997.
[23] W. Ritz, *Ann. Chim. Phys.* **13**, 145 (1908); translated and published by R. S. Fritzius, (1980).
[24] L. D. Landau and E. M. Lifshitz, *The Classical Theory of Fields* (Pergamon, Oxford, 1975).
[25] Using the relativistic Hamilton-Jacobi equation in Ref. 21, with elementary actions.

[26] B. Talin, V. P. Kaftandjan and L. Klein, *Phys. Rev. A* **11**, 648 (1975).

[27] P. W. Atkins, *Molecular Quantum Mechanics*, (Oxford University Press., 1983).

Paper 2

On the Use of a Complex Vector Potential in the Minimal Prescription in the Dirac Equation

It is argued that the use of a complex vector potential in the minimal prescription maintains the basic Hermitian property of the Hamiltonian operator in Dirac's equation for the interaction of a fermion with the classical electromagnetic field. This is demonstrated by setting up the Dirac equation for a complex vector potential and for its complex conjugate, then forming a pure real Hamiltonian.

Key words: Minimal prescription; Dirac equation; complex vector potential.

2.1 Introduction

In the standard theory [1] the minimal prescription is used in the Dirac equation with a real vector potential. This method reproduces the standard description of the Stern-Gerlach experiment but does not allow for a coupling between the conjugate product [2—6] of the electromagnetic field and the Pauli matrix of the fermion. In this paper we show that the use of a complex vector potential in the minimal prescription results in a pure real Hamiltonian operator which maintains the Hermitian properties of the Dirac equation intact *and* gives a direct coupling between the conjugate product and the Pauli matrix σ. In Sec. 2.2 we set up the equations necessary for this demonstration and solve them in Sec. 2.3. A short discussion is given of the consequences to nuclear magnetic resonance and electron spin resonance of the existence of such a direct coupling.

2.2 The Dirac Equation with a Complex Vector Potential

The use of a complex vector potential to describe electromagnetic radiation is a mathematical procedure in classical electrodynamics in which it is understood that the physical part is the real part [7,8]. In quantum mechanics the eigenvalues of Hermitian operators are real [1,9] and the eigenstates corresponding to different eigenvalues of Hermitian operators are orthogonal. Therefore it seems to be assumed implicitly that physical eigenvalues generated by the Schrödinger or Dirac equation must be pure real in order to be physical. This is a different rule from the one used in classical electrodynamics. In this paper we consistently use the same rule for both classical and quantum mechanics, and assume that the real part of a complex eigenvalue is physical. Therefore we use a complex, multiplicative, vector operator A multiplying the wavefunction (eigenfunction) in the basic wave equation, in this case a Dirac equation.

Proceeding on the basis of this working hypothesis, two Dirac equations are written for the interaction of the fermion with the classical electromagnetic field, represented respectively with a complex $\boldsymbol{A}$ and its conjugate $\boldsymbol{A}^*$. The real parts of $\boldsymbol{A}$ and $\boldsymbol{A}^*$ are the same, so if we work on the basis of the rule that the real part of the complex operator is physical, the two Dirac equations become the same. Therefore the use of $\boldsymbol{A}$ and $\boldsymbol{A}^*$ in this way is a working mathematical hypothesis, one which leads to a pure real (and Hermitian) Hamiltonian operator. It is shown in Sec. 2.2 that this method leads to a well known result first given by Volkov [10] but in addition gives the interaction energy between the conjugate product $\boldsymbol{A} \times \boldsymbol{A}^*$

of the classical radiation field with the Pauli spinor σ. This interaction energy is proportional to intensity I divided by the square of angular frequency ω.

The Dirac equation is

$$H\psi = E\psi , \tag{2.2.1}$$

where the Hamiltonian operator is

$$H = c\alpha \cdot \left(\boldsymbol{p} - \frac{e}{c}\boldsymbol{A} \right) + \beta mc^2 + eV . \tag{2.2.2}$$

Here ψ is the four-component Dirac spinor [1], and E the energy eigenvalue. The physical energy eigenvalue is real, so it is assumed usually that $\boldsymbol{A}$ must be real. In Eq. (2.2.2) we use Gaussian units [11]. Here c is the velocity of light, $\boldsymbol{p}$ the assumed real momentum of the fermion, e its charge and m its mass. The scalar potential V is also assumed to be pure real in the conventional method [1,9]. The matrices α and β are defined by

$$\alpha = \begin{pmatrix} 0 & \sigma \\ \sigma & 0 \end{pmatrix}, \quad \beta = \begin{pmatrix} \mathbf{1} & 0 \\ 0 & \mathbf{1} \end{pmatrix}, \tag{2.2.3}$$

where σ is the Pauli matrix and $\mathbf{1}$ is the unit matrix. Following standard methods this equation is modified for the rest energy to

$$\left(E + mc^2\right)\psi' = \left(c\,\alpha \cdot \pi + \beta mc^2 + eV\right)\psi' , \tag{2.2.4}$$

where the modified spinor can be expressed as two two-component spinors, ψ_A and ψ_B. The minimal prescription is expressed through [11] the pure real,

$$\pi = \boldsymbol{p} - \frac{e}{c}\boldsymbol{A} , \tag{2.2.5}$$

and the Dirac equation splits into two interlinked equations,

$$\psi_B = \frac{c\sigma\cdot\pi}{E + 2mc^2 - eV}\psi_A, \tag{2.2.6}$$

$$(E - eV)\,\psi_A = \frac{c^2(\sigma\cdot\pi)^2}{E + 2mc^2 - eV}\psi_A. \tag{2.2.7}$$

One of these is an equation in ψ_A and the other links ψ_A to ψ_B. The second can be expressed as the wave equation,

$$H\psi_A = E\psi_A, \tag{2.2.8}$$

where H is the Hamiltonian,

$$H = \frac{c^2(\sigma\cdot\pi)^2}{E + 2mc^2 - eV} + eV. \tag{2.2.9}$$

This standard textbook procedure evidently gives a satisfactorily Hermitian equation which gives real and physical energy eigenvalues, positive and negative [1], but pure real.

If we let $\boldsymbol{A}$ be complex, we can write two Dirac equations,

$$(E - eV)\,\psi_A = c\sigma\cdot\pi\,\psi_B, \tag{2.2.10}$$

$$\left(E + 2mc^2 - eV\right)\psi_B = c\sigma\cdot\pi\,\psi_A, \tag{2.2.11}$$

$$(E - eV)\,\psi_A = c\,\sigma\cdot\pi^*\,\psi_B, \tag{2.2.12}$$

$$\left(E + 2mc^2 - eV\right)\psi_B = c\sigma\cdot\pi^*\,\psi_A, \tag{2.2.13}$$

and we can evaluate the consequences of this working hypothesis. Using Eqs. (2.2.10) and (2.2.13); or using (2.2.11) and (2.2.12) we obtain in both cases,

$$H\psi_A = E\psi_A, \tag{2.2.14}$$

$$H = \frac{c^2(\sigma\cdot\pi)(\sigma\cdot\pi^*)}{E + 2mc^2 - eV} + eV. \tag{2.2.15}$$

This equation is identical to Eq (26) of Ref. 10 except for a change of sign in V and a factor 2 multiplying mc^2 in the denominator. Using the standard [1] non-relativistic approximation,

$$E - eV \ll 2mc^2, \tag{2.2.16}$$

Eq. (2.2.15) can be written as,

$$H = H_1 + H_2 + \ldots, \tag{2.2.17}$$

$$H_1 := \frac{e^2}{2mc^2}\boldsymbol{A}\cdot\boldsymbol{A}^*, \tag{2.2.18}$$

$$H_2 := \frac{e^2}{2mc^2}\, i\sigma\cdot\boldsymbol{A}\times\boldsymbol{A}^*. \tag{2.2.19}$$

The term labeled H_1 was first derived [10] by Volkov in 1935 and is pure real. It is the second order contribution of the electromagnetic field to the kinetic energy of the particle in the non-relativistic limit. The dot product $\boldsymbol{A} \cdot \boldsymbol{A}^*$ is often referred to in classical electrodynamics as a time average over many cycles. Therefore the working hypothesis that $\boldsymbol{A}$ can be complex in Eqs. (2.2.14) and (2.2.15) leads to a standard Volkov result [10] for the theory of the Dirac fermion in the classical electromagnetic field.

The same hypothesis also leads to a novel term,

$$H_2 := \frac{e^2}{2mc^2} i\sigma \cdot \boldsymbol{A} \times \boldsymbol{A}^*, \tag{2.2.20}$$

which in *S.I.* units [10] becomes

$$H_2 = \frac{e^2}{2m} i\sigma \cdot \boldsymbol{A} \times \boldsymbol{A}^*, \tag{2.2.21}$$

and is also *pure real*. Thus, Volkov's term $\boldsymbol{H}_1$ is accompanied by the term H_2, which was first proposed using different methods in Ref 10.

2.3 The H_2 Term And Its Physical Meaning

The H_2 term represents a coupling between the half integral spin $\hbar\sigma / 2$ of the fermion and the conjugate product of the classical field, $\boldsymbol{A} \times \boldsymbol{A}^*$. It can be shown straightforwardly [10] that this is proportional to I/ω^2,

$$H_2 = \frac{e^2 c^2 B^{(0)2}}{2m\omega^2} i\sigma \cdot \boldsymbol{k}. \tag{2.2.22}$$

Therefore it allows the possibility of resonance between states of the spinor induced not by a magnet but by the conjugate product $\boldsymbol{A} \times \boldsymbol{A}^*$ [3—6]. The resonance frequency is given in *S.I.* units by,

$$f_{\text{res}} = \left(\frac{e^2 \mu_0 c}{2\pi\hbar m} \right) \frac{I}{\omega^2}, \tag{2.2.23}$$

where μ_0 is the vacuum permeability, and for a given *S.I.* increases as the inverse square of ω. As shown in Ref. 10 this property is potentially of great usefulness if developed experimentally. The theory also reproduces the order of magnitude of optical *NMR* shifts [12] introduced at visible frequencies.

2.4 Discussion

The Volkov term □$_1$ cannot be produced from the Dirac equation if we use the standard approach, that $\boldsymbol{A}$ is real. Yet it is a term which has made its way into a standard textbook such as that of Itzykson and Zuber [13]. The standard theory produces a term proportional to $\boldsymbol{A} \cdot \boldsymbol{A}$, which is highly oscillatory for electromagnetic radiation, and which is zero at high frequencies. Yet it is well known [14] that there exist non-linear optical effects proportional to the square of potential and field quantities. One of these is the inverse Faraday effect, which is static magnetization by a circularly polarized electromagnetic field, and which is described phenomenologically with $\boldsymbol{A} \times \boldsymbol{A}^*$ [15]. There are therefore internal inconsistencies in the standard fermion-field theory of the Dirac equation.

In the standard approach, in which $\boldsymbol{A}$ is pure real, the resonance frequency described by Eq. (2.2.23) becomes proportional to I/ω,

$$f_{\text{res}} = \left(\frac{\mu_0}{4\pi c} \frac{e^2}{m^2} \right) \frac{I}{\omega}, \tag{2.2.24}$$

and so it is easily possible in theory to test the working hypothesis on which is based Eqs. (2.2.10) to (2.2.13). If one is not allowed to use a complex A then the resonance frequency is proportional to I/ω ; otherwise it is proportional to I/ω^2. A simple beam experiment ought to be able to distinguish between these predictions experimentally, or to show that both are correct. In any event, the experimental demonstration of radiation induced fermion resonance would be of great practical value, and the theory of this effect has been developed elsewhere [16].

Acknowledgments

The Alpha Foundation is thanked for the award of a praesidium membership, honoris causa, and many colleagues for interesting discussions.

References

[1] L. H. Ryder, *Quantum Field Theory* (Cambridge University Press, Cambridge, 1987).

[2] M. W. Evans and S. Kielich, eds., *Modern Nonlinear Optics*, in *Advances in Chemical Physic*, I Prigogine and S. A. Rice, eds. (Wiley Interscience, New York, 1997.)

[3] W. Happer, *Rev. Mod. Phys.* **171**, 11 (1968).

[4] J. P. van der Ziel, P. S. Pershan and L. D. Malmstrom, *Phys. Rev. Lett.* **15**, 190 (1965); J. Deschamps, M. Fitaire and M. Lagoutte, *Phys. Rev. Lett.* **25**, 1330 (1970).

[5] G. H. Wagnière, *Linear and Nonlinear Optical Properties of Molecules* (VCH, Basel, 1993).

[6] P. S. Pershan, *Phys. Rev.* **130**, 919 (1963).

[7] J. D. Jackson, *Classical Electrodynamics* (Wiley, New York, 1962).

[8] W. K. H. Panofsky and M. Phillips, *Classical Electricity and Magnetism* 2nd. ed. (Addison-Wesley, Reading, Mass., 1962).

[9] P. W. Atkins, *Molecular Quantum Mechanics* (Oxford University Press, Oxford, 1983).

[10] M. W. Evans, J.-P. Vigier, S. Roy and S. Jeffers, *The Enigmatic Photon, Volume 3: Theory and Practice of the* $\boldsymbol{B}^{(3)}$ *Field.* (Kluwer, Dordrecht, 1996), Chaps. 1 and 2.

[11] J. D. Bjorken and S. D. Drell, *Relativistic Quantum Mechanics* (McGraw-Hill, New York, 1964).

[12] W. S. Warren, S. Mayr, D. Goswami and A. P. West, Jr., *Science* **255**, 1683 (1992); **259**, 836 (1993).

[13] C. Itzykson and J.-B. Zuber, *Quantum Field* Theory (McGraw Hill, New York, 1980), which discusses the Volkov term described in Ref. 10.

[14] G. Wagnière, *Phys. Rev. A* **40**, 2437 (1989).

[15] S. Woźniak, M. W. Evans and G. Wagnière, *Mol. Phys.* **75**, 81, 99 (1992).

[16] S. Esposito and M. W. Evans, *Phys. Rev. A* submitted for publication.

Paper 3

Infinitesimal Field Generators

The concept of infinitesimal field generator is introduced, using the principle that the underlying symmetry of special relativity is described by the Poincaré group, or ten generator inhomogeneous Lorentz group. This concept leads straightforwardly to the vacuum Maxwell equations and six cyclical relations between field components. Both the Maxwell equations and the field relations are given by the *E(2)* little group. The recently inferred *field spin*, the vacuum magnetic flux density labeled $\boldsymbol{B}^{(3)}$ in the complex basis ((1), (2), (3)), is rigorously non-zero from first principles. The existence of $\boldsymbol{B}^{(3)}$ is compatible with the vacuum Maxwell equations, and the cyclic relations are automatically covariant. The B cyclics are invariants of the classical field.

Key words: Infinitesimal field generators; $\boldsymbol{B}^{(3)}$ field; invariance of B cyclics.

3.1 Introduction

The concept is introduced at the classical level of infinitesimal generators of the vacuum electromagnetic field, and it is shown that the Maxwell equations can be deduced from this concept along with cyclical field relations involving the recently inferred *field spin* labeled $\boldsymbol{B}^{(3)}$ in the complex basis ((1), (2), (3)) [1—20]. It follows that the $\boldsymbol{B}^{(3)}$ field is rigorously non-zero if we accept our first principle, which asserts that the infinitesimal field generators are generators of the inhomogeneous Lorentz group. The antisymmetric matrix of field generators is interpreted in terms of intrinsic spin. The theory is automatically Lorentz covariant (independent of any frame of reference) and therefore compatible with the principle of relativity, and puts the theory of the classical electromagnetic field into close correspondence with Wigner's theory [21] of particles, a theory which ascribes to each particle an intrinsic mass and spin.

Section 3.2 describes the relativistic theory of spin angular momentum, and using this framework proposes an equivalence between the infinitesimal generators used to define this theory and novel infinitesimal generators of magnetic flux density and electric field strength in vacuo. The infinitesimal translation generator is made equivalent to the infinitesimal generator of a fully covariant vector potential. The Pauli-Lubanski operator of the relativistic spin angular momentum theory [22] is equivalent to the vector introduced by Afanasiev *et al.* [23] which is formed by multiplying the matrix of infinitesimal field operators with the infinitesimal generator of potential.

Section 3.3 is a straightforward deduction of the Lie algebra of field generators in the light-like condition. This algebra has the *E(2)* symmetry — that of a well known little group of the Poincaré group [22]. The *E(2)* symmetry is that of commutator relations between the novel infinitesimal field generators. Particular solutions of this Lie algebra are shown straightforwardly to be consistent with the vacuum Maxwell equations, which are thereby *deduced* from our first principle — that the underlying symmetry group of special relativity is the Poincaré group. The same *E(2)* Lie algebra gives six cyclical relations between eigenvalues of the field generators in the basis ((1), (2), (3)) [1—10]. Of these, three form the B

Cyclic theorem [1—10] in the light-like condition. It is easily shown that the B Cyclic theorem retains its form in the rest-frame (the hypothetical rest frame corresponding to a massive photon), and so the theorem is a Lorentz *invariant*. The algebra shows conclusively that $\boldsymbol{B}^{(3)}$ is rigorously non-zero from the first principle adopted for this development, i.e., that the Poincaré group is the symmetry group of special relativity. If one accepts this first principle, it follows that $\boldsymbol{B}^{(3)}$, the *field spin*, is rigorously non-zero. It is difficult to see how the Poincaré group cannot be the symmetry group of special relativity, and it is concluded that $\boldsymbol{B}^{(3)}$ is the fundamental spin of the classical electromagnetic field in vacuo. It has been shown recently [1—20] that $\boldsymbol{B}^{(3)}$ is only one of several possible longitudinal solutions in vacuo of the Maxwell equations, which as this work shows, are possible conservation equations compatible with the *E(2)* Lie algebra of infinitesimal field generators in the light-like condition. The antisymmetric matrix of generators is an intrinsic spin of the classical electromagnetic field, corresponding to the intrinsic spin angular momentum of the photon. The latter is described by the well known antisymmetric matrix $J_{\mu\nu}$ of relativistic angular momentum theory [22], and by the Pauli-Lubanski pseudo four- vector W^{μ} [22].

Finally we discuss this result in terms of a precise correspondence between intrinsic field spin and intrinsic photon spin in quantum field theory.

3.2 Relativistic Spin Angular Momentum Theory and Infinitesimal Field Generators

The theory of relativistic spin angular momentum for particles is developed for example by Barut [24] and Ryder [22] and relies on the Pauli-Lubanski pseudo four-vector. The latter is dual in four dimensions to the antisymmetric spin angular momentum tensor, and cannot be defined without the introduction of the energy-momentum four-vector. The Pauli-Lubanski four- vector is therefore,

$$W^{\lambda} := -\frac{1}{2}\epsilon^{\lambda\mu\nu\rho}p_{\mu}J_{\nu\rho}\,, \tag{2.3.1}$$

where $\epsilon^{\lambda\mu\nu\rho}$ (with $\epsilon^{0123} = 1$) is the antisymmetric unit four-tensor. The antisymmetric natrix $J_{\nu\rho}$ is given by,

$$J_{\nu\rho} = \begin{bmatrix} 0 & K_1 & K_2 & K_3 \\ -K_1 & 0 & -J_3 & J_2 \\ -K_2 & J_3 & 0 & -J_1 \\ -K_3 & -J_2 & J_1 & 0 \end{bmatrix}, \tag{2.3.2}$$

where every element is an element of spin angular momentum in four dimensions. The energy-momentum four-vector is defined as usual by,

$$p^{\mu} = \left(p^0, \boldsymbol{p}\right) = \left(\frac{En}{c}, \boldsymbol{p}\right). \tag{2.3.3}$$

The Pauli-Lubanski pseudo four-vector is therefore a four dimensional cross product of angular and linear momentum for a classical particle.

It is well known [23,24] that these considerations can be extended to operators, infinitesimal generators of the Poincaré group (ten parameter inhomogeneous Lorentz group). In the operator representation $J_{\nu\rho}$ becomes a matrix of infinitesimal generators, of rotation generators J and boost generators K. The p_{μ} vector becomes the infinitesimal generator of translation in four dimensions [22,24]. The infinitesimal generators can be represented as matrices or as differential operator combinations [22]. The Pauli-Lubanski operator (W^{μ}) then becomes a product of the $J_{\nu\rho}$ and p_{μ} operators. Barut [24] shows that the Lie algebra of the W^{μ} operators is,

$$\left[W^{\mu}, W^{\nu}\right] = -i\epsilon^{\mu\nu\sigma\rho} p_{\sigma} W_{\rho}, \tag{2.3.4}$$

which is a four dimensional commutator relation. The theory is automatically covariant and therefore compatible with the principle of special relativity, that the laws of physics are frame independent. Equation

(2.3.4) gives the Lie algebra of intrinsic spin angular momentum, because rotation generators are angular momentum operators within a factor $\hbar$ [22]. Similarly, translation generators are energy momentum operators within a factor $\hbar$. This development leads to a straightforward particle interpretation after quantization and to Wigner's famous result that every particle has intrinsic spin, including the photon.

Our basic ansatz is to assume that this theory applies to the vacuum electromagnetic field, considered as a physical entity of space-time in the theory of special relativity. The intrinsic spin of the classical electromagnetic field is the magnetic flux density $\boldsymbol{B}^{(3)}$ [1—20]. Infinitesimal generators of rotation correspond with those of intrinsic magnetic flux density in vacuo; those of boost with intrinsic electric field strength; those of translation with intrinsic, fully covariant, field potential. Thus, the symbols are transmuted as follows,

$$J \rightarrow B, \qquad K \rightarrow E, \qquad P \rightarrow A . \tag{2.3.5}$$

In Cartesian notation, the Pauli-Lubanski vector of the particle theory becomes a pseudo four-vector operator of the classical electromagnetic field in the vacuum,

$$W^{\lambda} = -\frac{1}{2} \epsilon^{\lambda\mu\nu\rho} A_{\mu} F_{\nu\rho} . \tag{2.3.6}$$

The Lie algebra (2.3.4) becomes a Lie algebra of the field.

3.3 The *E(2)* Lie Algebra of The Field

If it is assumed that the electromagnetic field propagates at c in vacuo, then we must consider the Lie algebra (2.3.4) in a light-like condition. The latter is satisfied by a choice of (Appendix 3A),

$$A^{\mu} := \left(A^{0}, A_{Z}\right), \qquad A^{0} = A_{Z} , \tag{2.3.7}$$

corresponding in the particle interpretation to the light like translation generator,

$$p^{\mu} := \left(p^{0}, p_{Z}\right), \qquad p^{0} = p_{Z} . \tag{2.3.8}$$

The Pauli-Lubanski pseudo-vector of the field in this condition is,

$$\begin{aligned} W^{\mu} &= \left(A_Z B_Z, A_Z E_Y + A_0 B_X, -A_Z E_X + A_0 B_Y, A_0 B_Z\right) \\ &= A_0\left(B_Z, E_Y + B_X, -E_X + B_Y, B_Z\right), \end{aligned} \tag{2.3.9}$$

and the Lie algebra (2.3.4) becomes

$$\left.\begin{aligned} \left[B_X + E_Y, B_Y - E_X\right] &= i\,(B_Z - B_Z)\,, \\ \left[B_Y - E_X, B_Z\right] &= i\,(B_X + E_Y)\,, \\ \left[B_Z, B_X + E_Y\right] &= i\,(B_Y - E_X)\,, \end{aligned}\right] \tag{2.3.10}$$

which has *E(2)* symmetry. In the particle interpretation Eqs. (2.3.9) and (2.3.10) correspond to,

$$W^{\mu} = \left(p_Z J_Z, p_Z K_Y + p_0 J_X, -p_Z K_X + p_0 J_Y, p_0 J_Z\right) \tag{2.3.11}$$

and

$$\left.\begin{aligned} \left[J_X + K_Y, J_Y - K_X\right] &= i\,(J_Z - J_Z) \\ \left[J_Y - K_X, J_Z\right] &= i\,(J_X + K_Y) \\ \left[J_Z, J_X - K_Y\right] &= i\,(J_Y - K_X) \end{aligned}\right\} . \tag{2.3.12}$$

In the hypothetical rest frame the field and particle Pauli-Lubanski vectors are respectively

$$W^{\mu} = \left(0, A_0 B_X, A_0 B_Y, A_0 B_Z\right), \tag{2.3.13}$$

and

$$W^{\mu} = \left(0, P_0 J_X, P_0 J_Y, P_0 J_Z\right), \tag{2.3.14}$$

and the rest frame Lie algebra for field and particle is respectively (normalized $B^{(0)} = 1$ units)

$$\left[B_X, B_Y\right] = iB_Z \qquad (et\ cyclicum)\ , \tag{2.3.15}$$

and

$$\left[J_X, J_Y\right] = iJ_Z\,, \qquad (et\ cyclicum)\,. \tag{2.3.16}$$

It is straightforward to show that the *E(2)* field Lie algebra (2.3.10) is compatible with the vacuum Maxwell equations written for eigenvalues of our novel infinitesimal field operators. This is demonstrated as follows.

A particular solution of the *E(2)*, or little group, Lie algebra (2.3.10) is given by equating infinitesimal field generators as follows,

$$B_Y = E_X, \qquad B_X = -E_Y\ . \tag{2.3.17}$$

It is assumed that the eigenfunction (ψ) operated upon by these infinitesimal field generators is such that the same relation (2.3.17) holds between eigenvalues of the field. In order for this to be true the eigenfunction must be the de Broglie eigenfunction, i.e., the phase of the classical electromagnetic field,

$$\psi = e^{i(\omega t - \kappa Z)} , \tag{2.3.18}$$

where ω is the frequency at instant t and κ the wavevector at point Z. This is demonstrated in Appendix 3B.

The relation (2.3.17) interpreted as one between eigenvalues is compatible with the plane wave solution of Maxwell's vacuum equations [1—10] for circular polarization, i.e.,

$$\boldsymbol{E}^{(1)} = \boldsymbol{E}^{(2)*} = \frac{E^{(0)}}{\sqrt{2}} (\boldsymbol{i} - i\boldsymbol{j}) e^{i(\omega t - \kappa Z)}, \tag{2.3.19}$$

$$\boldsymbol{B}^{(1)} = \boldsymbol{B}^{(2)*} = \frac{B^{(0)}}{\sqrt{2}} (i\boldsymbol{i} + \boldsymbol{j}) e^{i(\omega t - \kappa Z)} , \tag{2.3.20}$$

and this conveniently introduces the complex basis ((1), (2), (3)) defined by the unit vectors [1—10],

$$\left. \begin{array}{c} \boldsymbol{e}^{(1)} = \boldsymbol{e}^{(2)*} := \frac{1}{\sqrt{2}} (\boldsymbol{i} - i\boldsymbol{j}) \\ \boldsymbol{e}^{(3)} = \boldsymbol{e}^{(3)*} := \boldsymbol{k} \end{array} \right\} . \tag{2.3.20}$$

It is concluded that our basic ansatz is compatible with Maxwell's vacuum equations, which are one possible way of ensuring that Eq. (2.3.17) holds.

It follows that the same analysis can be applied to the particle interpretation, giving,

$$\partial_\mu J^{\mu\nu} = \partial_\mu \tilde{J}^{\mu\nu} = 0 , \tag{2.3.21}$$

in the vacuum. This is a possible conservation equation (simple relation between spins) which is compatible with the *E(2)* symmetry of the little group. In the particle interpretation this little group symmetry is the one given by considering the most general Lorentz transform that leaves the light-like vector (2.3.8) invariant.

It is concluded that the vacuum Maxwell equations for the field correspond with Eq. (2.3.21) for the particle, an equation which asserts that the spin angular momentum matrix is divergentless. In vector notation we obtain from Eqs. (2.3.17) to (2.3.21) the familiar *S.I.* equations,

$$\left.\begin{aligned} \nabla \cdot \boldsymbol{B} &= 0\,, \\ \nabla \times \boldsymbol{E} + \frac{\partial \boldsymbol{B}}{\partial t} &= \boldsymbol{0}\,, \\ \nabla \times \boldsymbol{B} - \frac{1}{c^2}\frac{\partial \boldsymbol{E}}{\partial t} &= 0\,, \\ \nabla \cdot \boldsymbol{E} &= 0\,, \end{aligned}\right\} \tag{2.3.22}$$

and the less familiar relation between eigenvalues of spin angular momentum in four dimensions,

$$\left.\begin{aligned} \nabla \cdot \boldsymbol{J} &= 0\,, \\ \nabla \times \boldsymbol{J} + \frac{\partial K}{\partial t} &= \boldsymbol{0}\,, \\ \nabla \times \boldsymbol{K} - \frac{\partial \boldsymbol{J}}{\partial t} &= 0\,, \\ \nabla \cdot \boldsymbol{K} &= 0\,. \end{aligned}\right\} \tag{2.3.23}$$

Another particular solution of the *E(2)* Lie algebra (2.3.10) is given by commutator relations, of which there are six in total. Three of these form the recently inferred B cyclic theorem [1—10] ($B^{(0)} = 1$ units),

$$\left.\begin{aligned} \left[B_X, B_Y\right] &= iB_Z, \\ \left[B_Y, B_Z\right] &= iB_X, \\ \left[B_Z, B_X\right] &= iB_Y, \end{aligned}\right\} \tag{2.3.24}$$

and the other three are

$$\left.\begin{aligned} \left[E_X, E_Y\right] &= -iB_Z, \\ \left[B_Z, E_X\right] &= iE_Y, \\ \left[E_Y, B_Z\right] &= iE_X. \end{aligned}\right\} \tag{2.3.25}$$

In the particle interpretation these are parts of the Lie algebra [22,24] of rotation and boost generators of the Poincaré group,

$$\left.\begin{aligned} \left[J_X, J_Y\right] &= iJ_Z, \\ \left[J_Y, J_Z\right] &= iJ_X, \\ \left[J_Z, J_X\right] &= iJ_Y, \end{aligned}\right\} \tag{2.3.26}$$

$$\left.\begin{aligned} \left[K_X, K_Y\right] &= -iJ_Z, \\ \left[J_Z, K_X\right] &= iK_Y, \\ \left[K_Y, J_Z\right] &= iK_X. \end{aligned}\right\} \tag{2.3.27}$$

Using the methods sketched in Appendix 3B, we obtain from the Lie algebra of generators the following *S.I.* unit cyclic relations between field eigenvalues,

$$\left.\begin{aligned} \boldsymbol{B}^{(1)} \times \boldsymbol{B}^{(2)} &= iB^{(0)}\boldsymbol{B}^{(3)*}\,, \\ \boldsymbol{B}^{(2)} \times \boldsymbol{B}^{(3)} &= iB^{(0)}\boldsymbol{B}^{(1)*}\,, \\ \boldsymbol{B}^{(3)} \times \boldsymbol{B}^{(1)} &= iB^{(0)}\boldsymbol{B}^{(2)*}\,, \end{aligned}\right\} \tag{2.3.28}$$

$$\left.\begin{aligned} \boldsymbol{E}^{(1)} \times \boldsymbol{E}^{(2)} &= ic^2B^{(0)}\boldsymbol{B}^{(3)*}\,, \\ \boldsymbol{B}^{(3)} \times \boldsymbol{E}^{(1)} &= icB^{(0)}\boldsymbol{E}^{(2)*}\,, \\ \boldsymbol{B}^{(3)} \times \boldsymbol{E}^{(2)} &= -icB^{(0)}\boldsymbol{E}^{(1)*}\,, \end{aligned}\right\} \tag{2.3.29}$$

where $\boldsymbol{B}^{(3)} = B^{(0)}\boldsymbol{e}^{(3)}$. Similarly in the particle interpretation, and switching from rotation generator to spin angular momentum we obtain,

$$\left.\begin{aligned} \boldsymbol{J}^{(1)} \times \boldsymbol{J}^{(2)} &= i\hbar\boldsymbol{J}^{(3)*}\,, \\ \boldsymbol{J}^{(2)} \times \boldsymbol{J}^{(3)} &= i\hbar\boldsymbol{J}^{(1)*}\,, \\ \boldsymbol{J}^{(3)} \times \boldsymbol{J}^{(1)} &= i\hbar\boldsymbol{J}^{(2)*}\,. \end{aligned}\right\} \tag{2.3.30}$$

In the latter set of relations, $\hbar$ is the quantum of spin angular momentum.

In the hypothetical rest frame, we obtain for field and particle respectively, Eqs. (2.3.28) and (2.3.30), i.e., there are no boost generators as expected for a rest frame. The latter is hypothetical because an object translating at c identically does not have a rest frame by definition. We must therefore imagine an object translating infinitesimally close to c in vacuo in order to be able to back transform into a rest frame. This object can be thought of in our development as the electromagnetic field concomitant to photon with mass. In our new analysis the field and photon become topologically the same thing.

It is concluded that the $\boldsymbol{B}^{(3)}$ component is identically non-zero, otherwise all the field components vanish in the Lie algebra (2.3.24). If we assume Eq. (2.3.17), and at the same time assume that $\boldsymbol{B}^{(3)}$ is zero, then the Pauli-Lubanski pseudo four-vector (2.3.9) vanishes for all A_0. Similarly in the particle interpretation if we assume the equivalent of Eq. (2.3.17) and assume that $\boldsymbol{J}^{(3)}$ is zero, the Pauli-Lubanski vector W^{μ} vanishes. This is contrary to the definition of the helicity of the photon [22]. Therefore for finite field helicity we need a finite $\boldsymbol{B}^{(3)}$. The latter result is also indicated experimentally in magneto-optics [1—10], which can be used to observe the product $\boldsymbol{B}^{(1)} \times \boldsymbol{B}^{(2)}$. If $\boldsymbol{B}^{(3)}$ were zero this product would be zero, contrary to experience.

Therefore finite electromagnetic field helicity requires a finite $\boldsymbol{B}^{(3)}$ field in the light like condition. In the hypothetical rest frame a zero $\boldsymbol{B}^{(3)}$ would mean a zero $\boldsymbol{B}^{(1)} = \boldsymbol{B}^{(2)*}$.

3.4 Discussion

The precise correspondence between field and photon interpretation of vacuum electromagnetism developed here indicates that *E(2)* symmetry does not imply that $\boldsymbol{B}^{(3)}$ is zero, any more than it implies $\boldsymbol{J}^{(3)} = \boldsymbol{0}$. The assertion $\boldsymbol{B}^{(3)} = \boldsymbol{0}$ is counter indicated by magneto-optical data, and the B cyclics (2.3.28) are Lorentz covariant, being part of a Lorentz covariant Lie algebra. Furthermore, if one assumes the particular solution (2.3.24) and (2.3.25), and uses in it the particular solution (2.3.17), we obtain from the three cyclics Eq. (2.3.25) the cyclics (2.3.24), i.e., we obtain,

$$\left.\begin{aligned} \left[B_Y, -B_X\right] &= iB_Z, \\ \left[B_Z, B_Y\right] &= -iB_X, \\ \left[B_Z, -B_X\right] &= -iB_Y. \end{aligned}\right\} \tag{2.3.31}$$

This is also the relation obtained in the hypothetical rest frame. Therefore the B cyclic theorem is a Lorentz *invariant* in the sense that it is the same in the rest frame and in the light-like condition.

This result can be checked by applying the Lorentz transformation rules for magnetic fields term by term, i.e., to $B^{(0)}$, $\boldsymbol{B}^{(1)}$, $\boldsymbol{B}^{(2)}$ and $\boldsymbol{B}^{(3)}$ by considering a Lorentz boost at c in Z. The term $\boldsymbol{B}^{(1)} \times \boldsymbol{B}^{(2)}$ is transformed into itself multiplied by an indeterminate (0/0) which from the Lie algebra considered above is unity. The term $\boldsymbol{B}^{(3)}$ is unchanged, and $B^{(0)}$ must therefore be unchanged if we take the indeterminate to be unity. In the quantum interpretation,

$$B^{(0)} = \frac{\hbar\omega^2}{ec^2} = \frac{(\hbar\omega)\,\omega}{ec^2} = \left(\frac{\omega^2}{ec^2}\right)\hbar, \tag{2.3.32}$$

where e is the quantum of charge and c the speed of light in vacuo. If we consider $\hbar\omega$ to transform as energy and ω to transform as frequency [6] then $B^{(0)}$ is invariant under a Lorentz boost in Z. Since $\hbar\omega$ is the quantum of energy then it transforms as energy. Therefore it is concluded that the B cyclic theorem is invariant under a boost at c in Z. It appears unchanged in the Lie algebra of the light-like condition and of the rest frame as discussed already. (For intermediary boosts, taking place at v from the hypothetical rest frame, the numerical value of $\boldsymbol{B}^{(1)} \times \boldsymbol{B}^{(2)}$ is unchanged, but it becomes a function of v. The product $iB^{(0)}\boldsymbol{B}^{(3)*}$ is invariant again.)

It is concluded that the B cyclic theorem in the field interpretation is a Lorentz invariant construct. The equivalent of this result in the particle interpretation for spin angular momentum is that the J cyclic theorem,

$$\boldsymbol{J}^{(1)} \times \boldsymbol{J}^{(2)} = i\hbar\boldsymbol{J}^{(3)*}, \tag{2.3.33}$$

is a Lorentz invariant. This is compatible with the fact that $\hbar$ is an invariant and that $\boldsymbol{J}^{(3)}$ is invariant to a boost in Z. Thus $\boldsymbol{J}^{(1)} \times \boldsymbol{J}^{(2)}$ is invariant.

It is concluded overall that the ansatz adopted in this work is compatible both with the vacuum Maxwell equations and with the recently inferred cyclic relations between field components in vacuo [1—20]. As a result of this ansatz the $\boldsymbol{B}^{(3)}$ component in the field interpretation is non-zero in the light-like condition and in the rest frame, and is a solution of Maxwell's equations in the vacuum. The B cyclic theorem is a Lorentz invariant, and the product $\boldsymbol{B}^{(1)} \times \boldsymbol{B}^{(2)}$ is an experimental observable [1—10]. In this representation, $\boldsymbol{B}^{(3)}$ is a phaseless and fundamental field spin, an intrinsic property of the field in the same way as $\boldsymbol{J}^{(3)}$ is an intrinsic property of the photon. The scalar $B^{(0)}$ for the field plays the role of $\hbar$ for the photon, and if $\hbar\omega$ transforms as energy and ω as frequency, is also a Lorentz invariant. It is incorrect to infer from the Lie algebra (2.3.10) that $\boldsymbol{B}^{(3)}$ must be zero for plane waves. For the latter we have the particular choice (2.3.17) and the algebra (2.3.10) reduces to,

$$i\,(B_Z - B_Z) = 0\,, \tag{2.3.34}$$

which does not indicate that B_Z is zero any more than the equivalent particle interpretation indicates that J_Z is zero. That B_Z is zero is therefore a wholly unwarranted assumption of the literature [22]. Vacuum electromagnetism is *not* purely transverse in nature, and this result has recently been shown in several different ways [1—20]. By using the Poincaré group for vacuum electromagnetism it becomes easier to unify field theory [1—20], this particular paper has introduced the notion of infinitesimal field generators and has shown that this ansatz is compatible both with the vacuum Maxwell equations and the B cyclic theorem. The latter is a fundamental theorem of fields which shows that transverse solutions are always accompanied by longitudinal solutions.

Acknowledgments

It is a pleasure to acknowledge numerous Internet discussions and to thank the Alpha Foundation of the Institute of Physics, Budapest, Hungary, for an honorary membership.

References

[1] M. W. Evans, *Physica B* **182**, 227, 237 (1992); *Found. Phys.* **24**, 1671 (1994).

[2] M. W. Evans and S. Kielich eds., *Modern Nonlinear Optics,* Vols. 85(1—3) of *Advances in Chemical Physics,* I. Prigogine and S. A. Rice, eds. (Wiley Interscience, New York, 1997, third printing, paperback); M. W. Evans, *The Photon's Magnetic Field.* (World Scientific, Singapore, 1992); A. A. Hasanein and M. W. Evans, *The Photomagneton in Quantum Field Theory.* (World Scientific, Singapore, 1994).

[3] M. W. Evans and J.-P. Vigier, *The Enigmatic Photon, Vol. 1: The Field* $\boldsymbol{B}^{(3)}$ (Kluwer Academic, Dordrecht, 1994).

[4] M. W. Evans and J.-P. Vigier, *The Enigmatic Photon, Vol. 2: Non-Abelian Electrodynamics* (Kluwer Academic, Dordrecht, 1995).

[5] M. W. Evans, J.-P. Vigier, S. Roy and S. Jeffers, *The Enigmatic Photon, Vol. 3:, Theory and Practice of the* $\boldsymbol{B}^{(3)}$ *Field.* (Kluwer, Dordrecht, 1996).

[6] M. W. Evans, J.-P. Vigier, and S. Roy, eds., *The Enigmatic Photon, Vol. 4: New Directions.* (Kluwer Academic, Dordrecht, 1998).

[7] M. W. Evans, *Physica A* **214**, 605 (1995); L. D. Barron, *Physica B* **190**, 307 (1993); M. W. Evans, ibid., p. 310; A. D. Buckingham and L. Parlett, *Science* **264**, 1748 (1994); A. D. Buckingham, *Science* **266**, 665 (1994); A. Lakhtakia, *Physica B* **191**, 362 (1993); D. M. Grimes, *Physica B* **191**, 367 (1993); M. W. Evans, *Found. Phys. Lett.* **8,** 563 (1995); A. Lakhtakia, *Found. Phys. Lett.* **8**, 183 (1995); M. W. Evans, *Found. Phys. Lett.* **8,** 187 (1995); S. J. van Enk, *Found. Phys. Lett.* **9**, 183 (1996); M. W. Evans, ibid., 191; G. L. J. A. Rikken, *Opt. Lett.* **20**, 846 (1995); M. W. Evans, *Found. Phys. Lett.* **9**, 61 (1996); E. Comay, *Chem. Phys. Lett.* **261**, 601 (1996); M. W. Evans and S. Jeffers, *Found. Phys. Lett.* **9**, 587 (1996).

[9] M. W. Evans, *J. Phys. Chem.* **95**, 2256 (1991); W. S. Warren, S. Mayr, D. Goswami and A. P. West, Jr., *Science* **255**, 1683 (1992); ibid., **259**, 836 (1993); R. A. Harris and I. Tinoco, Jr., *Science* **259**, 835 (1993); ibid., *J. Chem. Phys.* **101**, 9289 (1994); M. W. Evans, Chaps. 1—2 of Ref. 6.

[10] C. R. Keyes, M. W. Evans and J.-P. Vigier, eds., *Apeiron* special issue on the $\boldsymbol{B}^{(3)}$ field, Spring 1997; M. W. Evans and J.-P. Vigier, *Poincaré Group Electrodynamics* (World Scientific, 1998, in prep.).

[11] V. V. Dvoeglazov, *Int. J. Mod. Phys.* **34**, 2467 (1995); *Rev. Mex. Fis.*, **41**, 159 (1995); *Nuov. Cim. A,* **108,** 1467 (1995); *Helv. Phys. Acta*, in press; review in ref. (6);*Majorana Like Models in the Physics of Neutral Particles.*, ICTE 95, (1995).

[12] H. A. Munera, D. Buritica, O. Guzman and J. I. Vallejo, *Rev. Colomb., Fis.*, **27**, 215 (1995); H. A. Munera and O. Guzman, *Found. Phys. Lett.* in press; ibid., part B, Magnetic Potentials, Longitudinal Currents and Magnetic Properties of the Vacuum, All Implicit in Maxwell's Equations.; ibid., *Phys. Rev. Lett.* submitted, A Symmetric Formulation of Maxwell's Equations.

[13] A. E. Chubykalo and R. Smirnov-Rueda, *Phys. Rev. E* **53**, 5373 (1996); *Int. J. Mod. Phys.*, in press; review in Ref. 6; monograph in prep. for World Scientific on the convective displacement current; A. E. Chubykalo, R. Smirnov-Rueda and M. W. Evans, *Found. Phys. Lett.* in press.

[14] B. Lehnert, *Phys. Scripta* **53**, 204 (1996); *Optik* **99**, 113 (1995); *Spec. Sci. Tech.* **17,** 259, 267 (1994); in Ref. 6, a review; and B. Lehnert and S. Roy, monograph in preparation for World Scientific.

[15] M. Meszaros, P. Molnar, T. Borbely and Z. G. Esztegar, in Ref. 6, a review.

[16] S. Jeffers, M. Novikov and G. Hathaway, in S. Jeffers, S. Roy, J.-P. Vigier and G. Hunter, eds., *The Present Status of the Quantum Theory of Light*. (Kluwer Academic, Dordrecht, 1997), pp. 127-139.

[17] E. Recami and M. W. Evans, a review in Ref. 6, E. Gianetto, *Let. Nuovo Cim.*, **44**, 140 (1985); E. Majorana, *Nuovo Cim.*, **14**, 171 (1937) and folios in the Domus Galilaeana, Pisa.

[18] V. V. Dvoeglazov, M. W. Evans, J.-P. Vigier *et al.* , The Photon and the Poincaré Group, volumes five to nine of Ref. (6), planned.

[19] M. Israelit, *Magnetic Monopoles and Massive Photons in a Weyl Type Electrodynamics.* LANL Preprint 9611060 (1996); *Found. Phys.* submitted for publication.

[20] D. Roscoe, e mail communications in 1996 and 1997, Dept. of Applied Mathematics, University of Salford.

[21] E. P. Wigner, *Ann. Math.* **140**, 149 (1939); Y. S. Kim, in N. M. Atakishiyev, K. B. Wolf and T. H. Seligman, eds., *Proceedings of the Sixth Wigner Symposium* (World Scientific, 1996), opening address.
[22] L. H. Ryder, *Quantum Field Theory,* 2nd edn. (Cambridge University Press, Cambridge, 1987).
[23] G. N. Afanasiev and Yu. P. Stepanovsky, *Nuovo Cim.* **109A**, 271 (1996); G. N. Afanasiev, *J. Phys. A* **26**, 731 (1993); **27**, 2143 (1994).
[24] A. O. Barut, *Electrodynamics and Classical Theory of Fields and Particles* (Macmillan, New York, 1964).

Appendix 3A. Basics of Poincaré Group Electrodynamics

The basic ansatz used in the text is that there exists a field vector analogous to the Pauli-Lubanski vector of particle physics; a field vector defined by,

$$W^{\lambda} := \tilde{F}^{\lambda\mu} A_{\mu}, \tag{2.3A.1}$$

where $\tilde{F}^{\lambda\mu}$ is the dual of the antisymmetric field tensor. Without assumptions of any kind, this vector has components,

$$\begin{aligned} W^0 &= -B^1 A_1 - B^2 A_2 - B^3 A_3, \\ W^1 &= B^1 A_0 + E^3 A_2 - E^2 A_3, \\ W^2 &= B^2 A_0 - E^3 A_1 + E^1 A_3, \\ W^3 &= B^3 A_0 + E^2 A_1 - E^1 A_2. \end{aligned} \tag{2.3A.2}$$

If we assume: a) that for the transverse components,

$$\boldsymbol{B} = \nabla \times \boldsymbol{A} \tag{2.3A.3}$$

if, b) B and A are plane waves,

$$\begin{aligned} \boldsymbol{A} &= \frac{A^{(0)}}{\sqrt{2}} (\boldsymbol{i}i + \boldsymbol{j}) e^{i\phi}, \\ \boldsymbol{B} &= \frac{B^{(0)}}{\sqrt{2}} (\boldsymbol{i}i + \boldsymbol{j}) e^{i\phi}, \end{aligned} \tag{2.3A.4}$$

and: c) if the longitudinal $\boldsymbol{E}^{(3)}$ is zero, then Eqs. (2.3A.2) reduce to those used in the text, i.e.,

$$\begin{aligned} W_0 &= A_Z B_Z, \\ W_X &= A_0 B_X + A_Z E_Y, \\ W_Y &= A_0 B_Y - A_Z E_X, \\ W_Z &= A_0 B_Z. \end{aligned} \tag{2.3A.5}$$

These assumptions mean that

$$A^{\mu} := \left(A^0, 0, 0, A^3\right), \qquad A^0 = A^3, \tag{2.3A.6}$$

can be used as an ansatz. Conversely, use of this definition means that the transverse components are plane waves and for the transverse components $\boldsymbol{B} = \nabla \times \boldsymbol{A}$.

In the Coulomb gauge the vector W^{μ} vanishes, meaning that there is no correspondence between particle and field theory for the Coulomb gauge or traditional assumption of transversality. Our final result is,

$$W^{\mu} = A^0\left(B_Z, 0, 0, B_Z\right), \tag{2.3A.7}$$

which is compatible with the *E(2)* Lie algebra of the text and with the vacuum Maxwell equations; together with $\boldsymbol{B} = \nabla \times \boldsymbol{A}$ for transverse components and,

$$\boldsymbol{B}^{(3)*} = -i\frac{\kappa}{A^{(0)}}\boldsymbol{A}^{(1)} \times \boldsymbol{A}^{(2)}, \tag{2.3A.8}$$

for longitudinal ones. It is significant that the $K^{(3)}\left(= K_Z\right)$ generator does not appear in the Lie algebra *E(2)*. This does not mean that $\boldsymbol{E}^{(3)}$ is zero necessarily, but it does not play the same role as $\boldsymbol{B}^{(3)}$. The latter is the most fundamental field spin, i.e., intrinsic spin of the classical electromagnetic field.

Appendix 3B. Inference of The De Broglie Wavefunction From Eq (2.3.17)

Equation.(2.3.17) of the text equates differential operators, i.e., field generators. The B operator is directly proportional to the J operator, the E operator to the K operator. In the vacuum $E^{(0)} = cB^{(0)}$ in *S.I.* units. These operate on a function such that the eigenvalues are related in the same way as the operators. Define the B and E differential operators by,

$$B_X \psi := -iB^{(0)} \left(Y \frac{\partial}{\partial Z} - Z \frac{\partial}{\partial Y} \right) \psi ,$$

$$E_Y \psi := -iE^{(0)} \left(t \frac{\partial}{\partial X} + X \frac{\partial}{\partial t} \right) \psi , \tag{2.3B.1}$$

and it is clear that the wavefunction is

$$\psi = e^{i(\omega t - \kappa Z)} , \tag{2.3B.2}$$

where $\kappa = \omega / c$. This is the well known phase of the vacuum electromagnetic wave, known sometimes as the de Broglie wavefunction.

Appendix 3C. Commutators to Cyclics

In order to translate a Cartesian commutator relation such as

$$\left[B_X, B_Y\right] = iB^{(0)}B_Z, \tag{2.3C.1}$$

to a ((1), (2), (3)) basis vector equation such as,

$$\boldsymbol{B}^{(1)} \times \boldsymbol{B}^{(2)} = iB^{(0)}\boldsymbol{B}^{(3)*}, \tag{2.3C.2}$$

consider firstly the usual unit vector relation in the Cartesian frame,

$$\boldsymbol{i} \times \boldsymbol{j} = \boldsymbol{k}. \tag{2.3C.3}$$

The unit vector $\boldsymbol{i}$ for example is defined by

$$\boldsymbol{i} := u_x \boldsymbol{i}, \tag{2.3C.4}$$

where u_x is a rotation generator [22] in general a matrix component. Therefore,

$$u_X = i(J_X)_{YZ}. \tag{2.3C.5}$$

The cross product $\times \boldsymbol{j}$ therefore becomes a commutator of matrices,

$$\left[J_X, J_Y\right] = iJ_Z, \tag{2.3C.6}$$

i.e.,

$$\left[J_X, J_Y\right] :=$$

$$\frac{1}{i}\begin{bmatrix} 0 & 0 & 0 \\ 0 & 0 & 1 \\ 0 & -1 & 0 \end{bmatrix}\frac{1}{i}\begin{bmatrix} 0 & 0 & -1 \\ 0 & 0 & 0 \\ 1 & 0 & 0 \end{bmatrix}-\frac{1}{i}\begin{bmatrix} 0 & 0 & -1 \\ 0 & 0 & 0 \\ 1 & 0 & 0 \end{bmatrix}\frac{1}{i}\begin{bmatrix} 0 & 0 & 0 \\ 0 & 0 & 1 \\ 0 & -1 & 0 \end{bmatrix} \tag{2.3C.7}$$

$$= \begin{bmatrix} 0 & 1 & 0 \\ -1 & 0 & 0 \\ 0 & 0 & 0 \end{bmatrix} := iJ_Z .$$

This can be extended immediately to angular momentum operators and infinitesimal magnetic field generators. Thus, a commutator such as (2.3C.1) is equivalent to a vector cross product. If we write $B^{(0)}$ as the scalar magnitude of magnetic flux density, the commutator (2.3C.1) becomes the vector cross product,

$$(B^{(0)}\boldsymbol{i}) \times (B^{(0)})\boldsymbol{j} = B^{(0)}(B^{(0)}\boldsymbol{k}) , \tag{2.3C.8}$$

which can be written conveniently as,

$$(B_X B_Y)^{1/2}\boldsymbol{i} \times (B_X B_Y)^{1/2}\boldsymbol{j} = iB^{(0)}B_Z\boldsymbol{k} . \tag{2.3C.9}$$

However, the Cartesian basis can be extended to the circular basis using relations between unit vectors [1—10], so Eq. (2.3C.9) can be written in the circular basis as,

$$(B_X B_Y)^{1/2}\boldsymbol{e}^{(1)} \times (B_X B_Y)^{1/2}\boldsymbol{e}^{(2)} = iB^{(0)}B_Z\boldsymbol{e}^{(3)*} , \tag{2.3C.10}$$

which is equivalent to,

$$\boldsymbol{B}^{(1)} \times \boldsymbol{B}^{(2)} = iB^{(0)}\boldsymbol{B}^{(3)*} , \tag{2.3C.11}$$

where we define

$$\boldsymbol{B}^{(1)} := (B_X B_Y)^{1/2} \boldsymbol{e}^{(1)} = \boldsymbol{B}^{(2)*},$$
$$\boldsymbol{B}^{(3)*} := B_Z \boldsymbol{e}^{(3)*}. \tag{2.3C.12}$$

To complete the derivation we multiply both sides of Eq. (2.3C.11) by the phase factor $e^{i\phi}e^{-i\phi}$ to obtain the B Cyclic theorem [1-10]. The latter is equivalent therefore to a commutator relation between infinitesimal magnetic field generators. Similarly,

$$\left[E_X, E_Y\right] = ic^2 B^{(0)} B_Z, \tag{2.3C.13}$$

is equivalent to

$$\boldsymbol{E}^{(1)} \times \boldsymbol{E}^{(2)} = ic^2 B^{(0)} \boldsymbol{B}^{(3)*}, \tag{2.3C.14}$$

and so forth.

Paper 4

Note on Radio Frequency Induced N.M.R.

The existence of radio frequency induced nuclear magnetic resonance (*RF-NMR*) is indicated by the Dirac equation. The recent theory of Harris and Tinoco [1] will not produce the main proton resonance because it has missed the key term.

4.1 Note

Recently, Harris and Tinoco [1] have used a perturbation theory to assert that the experimental data by Warren *et al.* [2,3] on optical *NMR* (*ONMR*) are inconsistent with the received view. It is asserted that light intensity produces negligible shifts in *NMR* spectra. This theory fails to reproduce the data reported by Warren *et al.* [2,3]. However, the Harris and Tinoco theory [1] is incomplete: the interaction Hamiltonian in their Eq. (4) contains no first order interaction between the nuclear spin from the Dirac equation and the radiation's conjugate product. For this reason the theory falls short of the data by seven or eight orders of magnitude. Using the first order spin term the eigenvalue of the interaction energy between the electromagnetic field and fermion (e.g. a proton) becomes [4] in Dirac's approximation [7],

$$W \sim \frac{e^2}{2m}\left(\boldsymbol{A}\cdot\boldsymbol{A}^* + i\sigma\cdot\boldsymbol{A}\times\boldsymbol{A}^* + \ldots\right), \tag{2.4.1}$$

where e/m is the charge to mass ratio of the fermion. Here $\boldsymbol{A}\times\boldsymbol{A}^*$ is the conjugate product of complex vector potentials [4—6] observed empirically in the inverse Faraday effect. The Pauli matrix σ forms an interaction energy with $\boldsymbol{A}\times\boldsymbol{A}^*$ from the Dirac equation [4] and this term is missed by Harris and Tinoco [1]. Resonance occurs between the two topological states of the spinor as in ordinary *NMR*. A simple calculation [4] shows that the probe resonance angular frequency for a proton is

$$\omega_{\text{res}}\left({}^1H\right) = 1.532\times 10^{25}\frac{I}{\omega^2}, \tag{2.4.2}$$

where ω is the pump angular frequency and I its intensity (watts per unit area). Superimposed on this main resonance (that of the bare proton unshielded by electrons) is the most useful feature of *RF-NMR*, the chemical shift spectrum [8]. The first term of our Eq. (2.4.1) is, within a factor $1/c^2$, the first term of Eq. (4) of Ref. 1. The second, spinor, term of our Eq. (2.4.1) is missing from Eq. (4) of Ref. 1 because Harris and Tinoco did not consider the direct interaction between the conjugate product [4—6] $\boldsymbol{A}\times\boldsymbol{A}^*$ and σ. The I/ω^2 coefficient of our Eq. (2.4.2) also appears in the top line, second column, page 9291, of Ref. 1, premultipied by a factor $2\pi c\epsilon\epsilon^*$. Thus Harris and Tinoco confirm our result [4—6] that $A^2 \propto I/\omega^2$, the key to *RF-NMR*.

The Ar^+ laser frequencies reported by Goswami [3] are 528.7, 488 and 476.5 *nm*. Taking I to be 10 watts per square centimeter we find probe resonance frequencies from Eq. (2.4.2) of 0.12, 0.10 and 0.09(8) *Hz* respectively. These are the main unshielded proton resonances and are of the same order of magnitude as the experimental data [2,3], obtained at the extreme edge of what is possible with contemporary laser technology. If the pump frequency is reduced however to the radio frequency range the main probe resonance frequency should appear from Eq. (2.4.2) in the infra red to visible region [4] for constant I of 10 watts per square centimeter. *This is* *an indication of the Dirac equation itself.* Harris and Tinoco [1] calculated minute second order chemical shift changes using a perturbation theory applied to the shielding constant, missing the main mechanism of resonance. Clearly, Eq. (2.4.2) indicates a major advance in *NMR* technology if implemented in the laboratory, removing the need for super-conducting magnets and producing very high resolution *NMR* in the infra-red and visible regions of the spectrum. If these features are not observed experimentally the Dirac equation would have failed. This hypothetical (and improbable) failure would have nothing to do with $\boldsymbol{B}^{(3)}$ theory [4—6] however, because Eq. (2.4.1) uses only $\boldsymbol{A} \times \boldsymbol{A}^*$, a property which has been verified empirically in the inverse Faraday effect [4—6], and which also used by Harris and Tinoco to calculate light induced shifts in chemical shifts [1].

It is of the utmost practical importance to realize that even if we accept uncritically the small, second order, light shift of 10^{-7} *Hz* estimated in Ref. 1, this is increased to *no less than 10 MHz* if the light frequency is reduced from visible (order of 10^{15} *Hz*) to radio frequency (order of 10^{8} *Hz*) for the same intensity. This alone, if realized empirically, would change all *NMR* and associated technology out of recognition: the resolution of the chemical shift would be enhanced enormously. It cannot be gainsaid, however, that Harris and Tinoco [1] have missed the main first order mechanism, one which if realized empirically will allow nuclear magnetic resonance spectra to be obtained routinely *in the visible range of frequencies without the use of magnets.* This would be of immense potential benefit to science and medicine.

Acknowledgments

York University, Toronto; and the Indian Statistical Institute are thanked for visiting professorships.

References

[1] R. A. Harris and I. Tinoco, Jr., *J. Phys. Chem.* **101(11)**, 9289 (1994); also A. D. Buckingham and L. L. Parlett, *Science* **264**, 1748 (1994).

[2] W. S. Warren, S. Mayr, D. Goswami, and A. P. West, Jr., *Science* **255**, 1683 (1992); also M. W. Evans, *J. Phys. Chem.* **95**, 2256 (1991).

[3] D. Goswami, Ph. D. Thesis (Princeton University, 1994).

[4] M. W. Evans, J.-P. Vigier, S. Roy, and S. Jeffers, *The Enigmatic Photon, Vol. 3: Theory and Practice of the* $\boldsymbol{B}^{(3)}$ *Field* (Kluwer Academic, Dordrecht, 1996), Chaps. 1 and 2.

[5] M. W. Evans and J.-P. Vigier, *The Enigmatic Photon, Vol. 1: The Field* $\boldsymbol{B}^{(3)}$ (Kluwer Academic, Dordrecht, 1994).

[6] M. W. Evans and J.-P. Vigier, *The Enigmatic Photon, Vol. 2: Non-Abelian Electrodynamics* (Kluwer Academic, Dordrecht, 1995).

[7] P. A. M. Dirac, *Quantum Mechanics*, 4th edn., (Oxford University Press, Oxford, 1974).

[8] P. W. Atkins, *Molecular Quantum Mechanics*, 2nd edn., (Oxford University Press, Oxford, 1983).

Paper 5

Fundamental Definitions for the Vacuum $\boldsymbol{B}^{(3)}$ Field

The fundamental definitions of the vacuum $\boldsymbol{B}^{(3)}$ field are developed in terms of the universal constants and radiation properties. The vacuum $\boldsymbol{B}^{(3)}$ field is the expectation value of the photomagneton operator $\hat{\boldsymbol{B}}^{(3)}$, an irremovable and fundamental property of the vacuum electromagnetic field.

5.1 Introduction

In the received view of electromagnetism in vacuo [1—3], the fields are transverse to the direction of propagation, and the photon is massless. Recently, this view has been challenged at the fundamental level by the proposal of the $\boldsymbol{B}^{(3)}$ (longitudinal) component, generated by the conjugate product of the transverse fields, a component which is phase free [4—10]. The existence of $\boldsymbol{B}^{(3)}$ is shown by the class of inverse Faraday induction phenomena [11—16], typified by the inverse Faraday effect, magnetization by radiation. Further experimental support for its existence would become available from the $I^{1/2}$ induction profile expected from $\boldsymbol{B}^{(3)}$ [17] at radio frequencies. Here I is the beam intensity in watts m^{-2}.

In this note, fundamental definitions of the vacuum $\boldsymbol{B}^{(3)}$ field are developed in terms of the universal constants and of fundamental radiation properties. The $\boldsymbol{B}^{(3)}$ field is defined as the irremovable and phase free expectation value of the photomagneton operator $\hat{\boldsymbol{B}}^{(3)}$ of one photon of energy $\hbar\omega$, where $\hbar$ is Dirac's constant and where ω is the angular frequency. This inference has recently been confirmed [18] by Muñera and Guzmán, who have shown the existence of a new class of longitudinal solutions in vacuo of the Maxwell equations. These authors isolated a component of their novel solutions which is phase independent and irremovable, thus confirming the earlier inference [4] that the photomagneton is a novel fundamental property of the photon and electromagnetic wave.

The monographs now available on $\boldsymbol{B}^{(3)}$ theory develop didactically the earliest theory [4—6], and clarify several aspects, linking up with work such as that of Hunter and Wadlinger [19] and Moles and Vigier [20]. It has been shown that the $\boldsymbol{B}^{(3)}$ field is defined in the vacuum by a component product of vector potentials $\boldsymbol{A}^{(1)} = \boldsymbol{A}^{(2)*}$,

$$\boldsymbol{B}^{(3)*} = -i\frac{e}{\hbar}\boldsymbol{A}^{(1)} \times \boldsymbol{A}^{(2)}, \tag{2.5.1}$$

where e is the charge quantum [4—10]. The mode of interaction of $\boldsymbol{B}^{(3)}$ with a fermion is determined by this definition through the Dirac equation. It is significant that $\boldsymbol{A}^{(1)} \times \boldsymbol{A}^{(2)}$ emerges from the Dirac equation itself [7] and is no longer phenomenological, as in the earliest papers [4—6]. The Dirac-Pauli and Hamilton-Jacobi equations can therefore be used to show the expected $I^{1/2}$ dependence of inverse induction due to $\boldsymbol{B}^{(3)}$; providing a route to empirical detection of $\boldsymbol{B}^{(3)}$ at first order. In general, inverse Faraday induction belongs to the class of non-linear optical phenomena [11—16], and depends on a non-linear optical property [4—10],

$$\boldsymbol{B}^{(1)} \times \boldsymbol{B}^{(2)} = iB^{(0)}\boldsymbol{B}^{(3)*}, \text{ et cyclicum,} \tag{2.5.2}$$

where $\boldsymbol{B}^{(1)} = \boldsymbol{B}^{(2)*}$ is the transverse magnetic component of, for example, a plane wave, in the circular basis ((1), (2), (3)). In Eq. (2.5.2), $B^{(0)}$ is the scalar magnitude of $\boldsymbol{B}^{(1)}$, and $\boldsymbol{B}^{(3)}$ is longitudinal and phase free. Experimental detection of $\boldsymbol{B}^{(3)}$ can therefore be achieved by showing the existence of the left hand side in Eq. (2.5.2). Effectively, this demonstration has been carried out, with the wisdom of hindsight, several times [11—16]. In the pre-1992 view, however, the existence of $\boldsymbol{B}^{(3)}$ through Eq. (2.5.2) was unknown, and the left hand side was constructed phenomenologically and known as the *conjugate product*.

Equation (2.5.1) has been developed recently within the framework of general relativity, using the inference [19] that the vacuum plane wave has a scalar curvature R, which in special relativity is not considered. (Curvatures and affine connections in Galilean space-time are zero by definition.) If the world-line of the charge quantum e is regarded as the fiducial geodesic in general relativity (a geodesic whose spatial trajectory is helical [19]) then Eq. (2.5.1) emerges from the Riemann tensor's antisymmetric contraction [7], giving a new equivalence principle for electromagnetism. If $\boldsymbol{B}^{(3)}$ is not considered, then a rigorously non-zero part of the Riemann curvature tensor disappears, the part that is quadratic in the affine connection. This inference opens new doors in field unification.

5.2 Fundamental Definitions in *S.I.* Units

In *S.I.* Units, the vacuum permeability is [21],

$$\mu_0 = 4\pi \times 10^{-7} Js^2C^{-2}m^{-1}, \tag{2.5.3}$$

and beam intensity, or power density, is measured as

$$I = \frac{c}{\mu_0} B^{(0)2} = cU_V, \tag{2.5.4}$$

where $B^{(0)} = |\boldsymbol{B}^{(3)}|$ and U_V is radiation energy per unit volume (Jm^{-3}). The units of $B^{(0)}$ are tesla (T) = $Wb\,m^{-2} = JsC^{-1}m^{-2}$. For the conventionally massless photon, c is the speed of light in ms^{-1}. Therefore the magnitude of the photomagneton is a magnetic flux (Wb) per unit area. From Eq. (2.5.4),

$$B^{(0)} = \left(\frac{\mu_0}{c}I\right)^{1/2} = (\mu_0 U_V)^{1/2}, \tag{2.5.5}$$

and is a conserved quantity in vacuo, being directly proportional to the *square root* of beam intensity. Under the right conditions [4—10], inverse induction due to $\boldsymbol{B}^{(3)}$ is also proportional to $I^{1/2}$, revealing the existence of $\boldsymbol{B}^{(3)} = B^{(0)}\boldsymbol{e}^{(3)}$.Here, $\boldsymbol{e}^{(3)}$ is a unit vector in the direction of propagation of the beam.

From Eq. (2.5.1),

$$B^{(0)} = \frac{e}{\hbar}A^{(0)2}, \tag{2.5.6}$$

where $A^{(0)} = |\boldsymbol{A}^{(1)}| = (\boldsymbol{A}^{(1)} \cdot \boldsymbol{A}^{(2)})^{1/2}$ in $JsC^{-1}m^{-1}$. Therefore,

$$A^{(0)2} = \left(\left(\frac{\mu_0}{c}\right)^{1/2}\frac{\hbar}{e}\right)I^{1/2}. \tag{2.5.7}$$

It is also known that $A^{(0)}$ and $B^{(0)}$ are related by the Maxwellian definition [21] of $\boldsymbol{A}$, i.e., $\boldsymbol{B} = \nabla \times \boldsymbol{A}$; and if $\boldsymbol{A}$ is taken to be a plane wave solution of the d'Alembert equation in vacuo, it follows [4—6] that

$$B^{(0)} = \kappa A^{(0)}, \tag{2.5.8}$$

where κ is the wavevector. For the conventional massless photon; $\kappa = \omega/c$. From Eqs. (2.5.6) and (2.5.8) emerges *the minimal prescription for the free photon* [4—10].

$$eA^{(0)} = \hbar\kappa, \tag{2.5.9}$$

an equation which balances the classical momentum per photon $eA^{(0)}$, with its quantum equivalent $\hbar\kappa$. In terms of the photon energy (the quantum of energy),

$$\hbar\omega = ecA^{(0)}. \tag{2.5.10}$$

For the sake of argument, we have accepted the idea of a massless photon in deriving Eq. (2.5.10) from Eq. (2.5.9). In contravariant notation, Eqs. (2.5.9) and (2.5.10) imply that the momentum/energy of the free photon is,

$$\begin{aligned} p^{\mu} &:= \left(En/c, \boldsymbol{p}\right) = \hbar\kappa^{\mu} := \left(\hbar\kappa, \hbar\boldsymbol{\kappa}\right) \\ &= eA^{\mu} := e\left(A^{(0)}, \boldsymbol{A}\right). \end{aligned} \tag{2.5.11}$$

Note that the relativistically correct result in Eq. (2.5.11) is incompatible with the transverse gauge [22], in which it is *assumed* that vacuum solutions of the d'Alembert equation have no longitudinal or time-like components. As shown by Muñera and Guzmán [18], *this assumption is incorrect*, there exists a class of longitudinal solutions under well-defined conditions more general than that of the transverse gauge. Therefore, as shown experimentally in the Aharonov-Bohm effects [23], A^{μ} is a physical observable, not a mathematical convenience. Within a factor e, A^{μ} is simply the energy momentum p^{μ} of the free photon, a gauge *invariant* physical observable. This in turn suggests that there is the need for a wave equation in the vacuum which restricts gauge freedom. An example is the Proca equation [4], which uses a very small, but non-zero, photon mass on which currently available experimental data put an upper bound [24]. Since massless particles conventionally [25] have only transverse degrees of polarization, the Proca equation is also implied by and compatible with $\boldsymbol{B}^{(3)}$ [4—10].

From Eq. (2.5.4), the energy per unit volume for one photon (the quantum of electromagnetic energy, $\hbar\omega$) is,

$$\frac{\hbar\omega}{V} = \frac{1}{\mu_0} B^{(0)2}, \tag{2.5.12}$$

where $B^{(0)}$ is the magnitude of the photomagneton, the quantum of magnetic flux density, and V is the average volume occupied by the photon. As shown by Hunter and Wadlinger [19], this is in general the volume of an ellipsoid, and in order to define this volume, the photon can be considered to be a wavicle, and not a particle. We therefore have three equations linking $A^{(0)}$ and $B^{(0)}$,

$$B^{(0)} = \frac{e}{\hbar} A^{(0)2}, \tag{2.5.13}$$

$$B^{(0)} = \frac{\omega}{c} A^{(0)}, \tag{2.5.14}$$

$$B^{(0)2} = \frac{\mu_0 e c}{V} A^{(0)}, \tag{2.5.15}$$

revealing the intricate inter-relations of basic vacuum electrodynamics.

A number of fundamental relations can now be derived from these three equations, in which $\hbar$, μ_0, c and e are universal constants and in which $A^{(0)}$, $B^{(0)}$, ω and V are electrodynamic quantities. From Eq. (2.5.13) in (2.5.15),

$$A^{(0)} = \frac{e\mu_0 c^3}{V\omega^2}, \tag{2.5.16a}$$

$$B^{(0)} = \frac{e\mu_0 c^2}{V\omega}. \tag{2.5.16b}$$

From Eq. (2.5.15) in (2.5.16b)

$$B^{(0)3} = \frac{e\mu_0^2 c^2 \hbar}{V^2}. \tag{2.5.17}$$

From Eq. (2.5.13) in (2.5.15),

$$A^{(0)} B^{(0)} = \frac{\mu_0 \hbar c}{V}. \tag{2.5.18}$$

From Eq. (2.5.14) in (2.5.18),

$$A^{(0)3} = \frac{\mu_0 c \hbar^2}{e} \frac{1}{V}. \tag{2.5.19}$$

From Eq. (2.5.16a) in (2.5.19) and (2.5.16b) in (2.5.17),

$$V = \frac{e^2 \mu_0 c^4}{\hbar} \frac{1}{\omega^3} . \qquad (2.5.20)$$

Throughout these equations $A^{(0)}$ appears as a physical quantity, not a supplementary mathematical variable. For example, Eq. (2.5.19) shows that $A^{(0)3}$ is inversely proportional to the volume V occupied by the photon of energy $\hbar\omega$, and $A^{(0)3}$ in consequence is as *physical* as V. An even more striking illustration of the physical nature of $A^{(0)}$ emerges from combining Eqs. (2.5.19) and (2.5.20) to give Eq. (2.5.10), which shows that $A^{(0)}$ is directly proportional to the observable ω.

Equation (2..5.20) shows that for any finite frequency, V is non-zero, meaning that the photon must always occupy a finite, frequency dependent, volume. It is a *point particle* only when ω is infinite, and at low enough frequencies, the volume V becomes macroscopic (e.g. order of km^3). There are obvious difficulties in continuing to accept the picture of a photon as an elementary particle of nuclear dimensions, for example. These have been carefully discussed by Hunter and Wadlinger [19], who also report experimental data on the finite volume of the photon as wavicle rather than particle. It is well known that de Broglie and Einstein attacked these difficulties using the empty wave hypothesis [24] and by locating all of the mass of the photon near its core, the rest being wave-like in nature. The received view [22] prohibits photon mass, so that at low enough frequencies we are asked to accept the existence of the photon as an elementary particle with no mass, but with macroscopic dimensions. Experiments on the radius and volume of the photon [19] should surely be used to test this counter-intuitive view.

An insight to the physical meaning of the relations between $B^{(0)}$ and $A^{(0)}$, Eqs. (2.5.13) to (2.5.15), can be obtained from the fact that $ecA^{(0)}$ has the dimensions of J (energy), and that $ec^2B^{(0)}$ has the dimensions of J s^{-1} or W (power). The latter is dimensionally the product of energy ($ecA^{(0)}$) and frequency (ω). Therefore $A^{(0)} = cB^{(0)}/\omega$ follows from the fact that energy is power divided by frequency. Equations (2.5.17) and (2.5.20) confirm that $B^{(0)}$ is proportional to *the quantum of power*, $\hbar\omega^2$,

$$B^{(0)} = \frac{\hbar\omega^2}{ec^2} = \frac{\hbar}{e}\kappa^2 . \tag{2.5.21}$$

The intensity equivalent to Eq. (2.5.21) gives *the radiation law for one photon*

$$I = \left(\frac{\hbar^2}{\mu_0 e^2 c^3}\right)\omega^4 . \tag{2.5.22}$$

This equation for one photon of energy $\hbar\omega$ is reminiscent of Stefan's law and Wien's law for black body radiation [21]; and it is deeply significant that these well known radiation laws stem from the fundamental relation between $B^{(0)}$ and V (Eq. (2.5.5)). In the last analysis the radiation laws emanate from the existence of V, the volume occupied by a photon of energy $\hbar\omega$. The depth of insight provided by this relation is revealed by considering the density of states of classical electromagnetic oscillators, as given by the Rayleigh-Jeans law [21],

$$\frac{dN}{d\nu} = \frac{8\pi\nu^2}{c^3}, \tag{2.5.23}$$

where N is the number of oscillators per m^3 and ν is the frequency ($\omega = 2\pi\nu$). The density of states is therefore,

$$\frac{dN}{d\nu} = \frac{2}{\pi c}\kappa^2 = \frac{2e}{\pi c\hbar}B^{(0)}, \tag{2.5.24}$$

where we have used Eq. (2.5.21). In terms of the scalar, or Gaussian, curvature, R [10] of the vacuum plane wave, we obtain,

$$\frac{dN}{d\nu} = \frac{2}{\pi c} R , \tag{2.5.25a}$$

$$B^{(0)} = \frac{\hbar}{e} R . \tag{2.5.25b}$$

In the generally relativistic theory of vacuum electromagnetism [10], R is the Gaussian curvature of the Riemann tensor (Sec. 5.1), *showing that the Rayleigh-Jeans density of states, and the photomagneton* $\boldsymbol{B}^{(3)}$ *are both manifestations of space-time curvature,* $\mathrm{R} = \kappa^2$. This inference allows radiation theory, notably the Planck distribution, to be developed as a theory of *general* relativity. At the most fundamental level, therefore, $\boldsymbol{B}^{(3)}$ is a property of curved space-time in general relativity, generated from the fact that the world-line of the charge quantum e is the fiducial geodesic. The trajectory of e in space is therefore a helix, and it becomes intuitively clear that this generates $\boldsymbol{B}^{(3)}$ along the axis of the helix (or *solenoid*).

The Planck distribution $\rho(\nu)$ is an expression for the mean energy $<\epsilon>$, of an electromagnetic oscillator of frequency ν when it can possess [21] only the discrete energies 0, $h\nu$, $2h\nu$, ..., $nh\nu$,

$$\begin{aligned}\rho(\nu) &= <\epsilon>\frac{dN}{d\nu} = \frac{2}{\pi c} R<\epsilon> = \frac{2e}{\pi c\hbar} B^{(0)}<\epsilon> \\ &= \frac{2}{\pi c} Rh\nu \left(\frac{e^{-h\nu/kT}}{1 - e^{-h\nu/kT}} \right) .\end{aligned} \tag{2.5.26}$$

5.3 Effect of Mass Density on Radiation Laws

The electromagnetic scalar curvature $R = \kappa^2$ appears therefore in the Planck distribution as a premultiplier. This is a scalar (or Gaussian) curvature in the theory of curvilinear coordinates [26], and Eq. (2.5.25b) demonstrates an *equivalence* between R and the field component $B^{(0)}$. The scalar curvature R_G from Einstein's equation [2] is, on the other hand,

$$R_G = -\frac{8\pi G}{c^2}\mu , \tag{2.5.27}$$

where μ is the mass density in kgm m^{-3}. Equation (2.5.27), in analogy with Eq. (2.5.25b), is an *equivalence* between R_G and μ, where G is the gravitational constant [2],

$$G = 6.67 \times 10^{-11}\, m^3\, \mathrm{kgm}^{-1}\, s^{-2} . \tag{2.5.28}$$

Both R and R_G are geometrical scalar curvatures in the theory of curvilinear coordinates, with the same units (m^{-2}). It is therefore logical to assume that electromagnetic and gravitational curvatures are additive, i.e., that R is changed to $R + R_G$ in the presence of mass density, μ. If it is assumed that such an effect does *not* exist, i.e., that electromagnetic and gravitational fields do not mix in this way, then a major philosophical fault-line develops, in that there exists an equivalence principle in gravitation, but none in electromagnetism, and that in consequence, $R = 0$, there is no curvature in the space-time of electromagnetism. However, R for a plane wave is κ^2, and is *not* zero. The received view treats electromagnetism [2] in a Galilean spacetime in which curvature is absent, but this clearly conflicts with $R = \kappa^2$; and recent work [4—10] has shown that it also conflicts with the existence of $\boldsymbol{B}^{(3)}$ because (Eq. (2.5.25b)) $\boldsymbol{B}^{(3)} = \Phi R \boldsymbol{e}^{(3)}$, where $\Phi = \hbar/e$

is the elementary fluxon (with the units of magnetic flux, weber) and where $\boldsymbol{e}^{(3)}$ is a unit vector in the propagation axis.

This section is therefore an attempt to develop simple cosmological tests for the hypothesis that R and R_G are additive. In direct logical consequence of this very simple ansatz, it can be shown as follows that the temperature and total photon density from a radiating black body are affected by its own mass density fluctuations. It may be possible to detect these small effects if the radiator is an object with a very large mass density μ, perhaps a neutron star, or the as yet unobserved black hole. In its simplest form the calculation assumes that in the presence of mass density, μ, the curvature of electromagnetism is changed by an amount determined from the Einstein equation,

$$\Delta R = (\Delta\kappa)^2 = \frac{8\pi G\Delta\mu}{c^2}, \tag{2.5.29}$$

so that the absolute change in electromagnetic frequency is proportional to the square root of the change in mass density,

$$\Delta\omega = (8\pi G)^{1/2}(\Delta\mu)^{1/2} \sim 4 \times 10^{-5}(\Delta\mu)^{1/2}. \tag{2.5.30}$$

It is now assumed that this frequency correction due to mass density fluctuation is the same for all electromagnetic frequencies in a radiating black body. Mass density fluctuations in the radiator therefore affect its own radiation properties such as radiated intensity and radiated photons per unit volume at a detector. The change in radiated energy per unit volume and photon density due to a change in μ of the black body radiator (e.g. a dense cosmic source) are, respectively, with,

$$\Delta\nu_0 = \frac{\Delta\omega_0}{2\pi} = \left(\frac{2G}{\pi}\right)^{1/2}(\Delta\mu)^{1/2}, \tag{2.5.31a}$$

$$\Delta U = \int_0^{\Delta\nu_0} \frac{8\pi h}{c^3}\nu^3\left(\frac{e^{-h\nu/kT}}{1-e^{-h\nu/kT}}\right)d\nu \sim \frac{8\pi kT}{3c^3}(\Delta\nu_0)^3, \tag{2.5.31b}$$

$$\Delta N = \int_0^{\Delta\nu_0} \frac{8\pi}{c^3}\nu^2\left(\frac{e^{-h\nu/kT}}{1-e^{-h\nu/kT}}\right)d\nu \sim \frac{2\pi kT}{hc^3}(\Delta\nu_0)^2, \tag{2.5.31c}$$

where we have used the classical approximation $h\nu/kT \ll 1$ [21]. The change in photon density for example is

$$\Delta N = \left(\frac{4kG}{hc^3}\right)T\Delta\mu \sim 10^{-25}\,T\Delta\mu \text{ photons } m^{-3}, \tag{2.5.32}$$

and is a small effect unless the product of T and mass density fluctuations in the radiator are very large. In these calculations k is Boltzmann's constant, and it is conjectured that there exists a physical upper bound on $\Delta\mu$; a physical mechanism which prevents the mass density of a radiating object from becoming infinite. Otherwise the electromagnetic frequency in Eq. (2.5.30) would also become infinite.

These small effects may be observable in solar physics with a very sensitive spectrometer with a sub-Hertzian resolution at visible frequencies. Although the solar mass is 1.989 x 10^{30} *kgm.*, the solar radius is 6.96 x 10^8 m; and the mean mass density is of the order 1000 kgm^{-3}; i.e about a gram per *c.c.* This seems too small to see the effects proposed here, but in general the mass density depends on the gravitational scalar potential and orbital parameters [2]. Data from different cosmic objects with very large mass densities must probably be used to test our ansatz that electromagnetic curvature adds to gravitational curvature within unified field theory. The existence of $\boldsymbol{B}^{(3)}$ [4—10] is already an experimental indication that electromagnetic and gravitational fields are both geometrical in origin through a general principle of equivalence, and as we have seen, $\boldsymbol{B}^{(3)}$ is proportional directly to R.

Acknowledgments

York University, Canada and the Indian Statistical Institute are thanked for the award of visiting professorships. Many interesting Internet discussions are gratefully acknowledged.

References

[1] J. D. Jackson, *Classical Electrodynamics* (Wiley, New York, 1962).

[2] L. D. Landau and E. M. Lifshitz, *The Classical Theory of Fields*, 4th edn. (Pergamon, Oxford, 1975).

[3] L. H. Ryder, *Quantum Field Theory*, 2nd edn. (Cambridge University Press, Cambridge, 1987).

[4] M. W. Evans, *Physica B* **182**, 227, 237 (1992); **183**, 103 (1993); *Physica A* **214**, 605 (1995).

[5] M. W. Evans, *The Photon's Magnetic Field* (World Scientific, Singapore, 1992).

[6] M. W. Evans and S. Kielich, eds., *Modern Nonlinear Optics,* Vols. 85(1), 85(2), 85(3) of *Advances in Chemical Physics,* I. Prigogine and S. A. Rice, eds., (Wiley Interscience, New York, 1993).

[7] M. W. Evans and J.-P. Vigier, *The Enigmatic Photon, Vol. 1: The Field* $\boldsymbol{B}^{(3)}$ (Kluwer Academic, Dordrecht, 1994).

[8] M. W. Evans and J.-P. Vigier, *The Enigmatic Photon, Vol. 2: Non-Abelian Electrodynamics* (Kluwer Academic, Dordrecht, 1995).

[9] M. W. Evans, J.-P. Vigier, S. Roy, and S. Jeffers, *The Enigmatic Photon, Vol. 3: Theory and Practice of the* $\boldsymbol{B}^{(3)}$ *Field* (Kluwer, Dordrecht, 1996).

[10] M. W. Evans, J.-P. Vigier, and S. Roy, eds. *The Enigmatic Photon, Vol. 4: General Relativity* (Kluwer, Dordrecht, in prep.); M. W. Evans and S. Roy, *Found. Phys.* in press; M. W. Evans, *Found. Phys. Lett.* **7**, 76, 209, 379, 467, 577 (1994); **8**, 63, 83, 187, 363, 385 (1995); *Found. Phys.* **24**, 892, 1519, 1671 (1994); **25**, 175, 383 (1995); A. A. Hasanein and M. W. Evans, *The Photomagneton in Quantum Field Theory* (World Scientific, Singapore, 1994).

[11] J.-P. van der Ziel, P. S. Pershan, and L. D. Malmstrom, *Phys. Rev. Lett.* 15, 190 (1965); *Phys. Rev.* **143**, 574 (1966).
[12] J. Deschamps, M. Fitaire, and M. Lagoutte, *Phys. Rev. Lett.* **25**, 1330 (1970); *Rev. Appl. Phys.* **7**, 155 (1972).
[13] W. Happer, *Rev. Mod. Phys.* **44**, 169 (1972).
[14] R. Zawodny, in Vol. 85(1) of Ref. 6, a review with ca. 140 refs.
[15] S. Woźniak, M. W. Evans, and G. Wagnière, *Mol. Phys.* 75, 81, 99 (1992).
[16] G. H. Wagniére, *Linear and Nonlinear Optical Properties of Molecules* (VCH, Basel, 1993).
[17] G. Hunter, S. Jeffers, S. Roy, and J.-P. Vigier, *Current Status of the Quantum Theory of Light* (Kluwer, Dordrecht, 1996), Proc. First Vigier Symposium, York University, Toronto, August 1995.
[18] H. A. Muñera and O. Guzmán, *Found. Phys. Lett.*, submitted.
[19] G. Hunter and R. L. P. Wadlinger, Phys. Essays, **2**, 156 (1989).
[20] M. Moles and J.-P. Vigier, *Comptes Rendues* **276**, 697 (1973).
[21] P. W. Atkins, *Molecular Quantum Mechanics*, 2nd edn. (Oxford University Press, Oxford, 1983).
[22] A. O. Barut, *Electrodynamics and the Classical Theory of Fields and Particles* (Macmillan, New York, 1964).
[23] M. W. Evans and S. Roy, *Found. Phys.*, submitted for publication.
[24] A. van der Merwe and A. Garuccio, eds., *Waves and Particles in Light and Matter* (Plenum, New York, 1994), Proc. de Broglie Centennial Symposium.
[25] E. P. Wigner, *Ann. Math.* **40**, 149 (1939).
[26] C. W. Misner, K. S. Thorne, and J. A. Wheeler, *Gravitation* (W. H. Freeman and Co., San Francisco, 1973).

Paper 6

Unified Field Theory and $\boldsymbol{B}^{(3)}$

The recent discovery of the vacuum spin field $\boldsymbol{B}^{(3)}$ of the electromagnetic sector means that unified field theory is also affected at the most fundamental level. It is shown that $\boldsymbol{B}^{(3)}$ changes the gauge symmetry of the electromagnetic sector from $U(1) = O(2)$ *to* $O(3)$, the rotation group symmetry. Accordingly, the massive bosons of *GWS* also acquire (3) components, but the ability of *GWS* to predict the correct masses is not affected.

Key words: $\boldsymbol{B}^{(3)}$ Field, Unified Field Theory.

6.1 Introduction

Electromagnetism in unified field theory [1—4] is conventionally the *U(1)* sector of theories such as *GWS* or *SU(5)*. The term *U(1) sector* derives from the $U(1) = O(2)$ gauge group that defines plane waves in the vacuum, the *O(2)* group being that of rotations in a plane, without reference to an orthogonal axis. In this conventional view, the physical fields are defined in the *O(2)* plane, and are transverse to the axis of propagation of the beam in the vacuum. Thus, for example, $\boldsymbol{B}^{(1)} = \boldsymbol{B}^{(2)*}$ is a plane wave of magnetic flux density that propagates in free space. By a careful examination of the conjugate product, $\boldsymbol{B}^{(1)} \times \boldsymbol{B}^{(2)}$, the conventional view just described has recently been changed fundamentally [5—12] because of the existence in the circular basis (1), (2), (3) of the Lie algebra:

$$\boldsymbol{B}^{(1)} \times \boldsymbol{B}^{(2)} = iB^{(0)}\boldsymbol{B}^{(3)*}, \qquad \boldsymbol{B}^{(2)} \times \boldsymbol{B}^{(3)} = iB^{(0)}\boldsymbol{B}^{(1)*},$$
$$\boldsymbol{B}^{(3)} \times \boldsymbol{B}^{(1)} = iB^{(0)}\boldsymbol{B}^{(2)*}, \tag{2.6.1}$$

where $\boldsymbol{B}^{(3)}$ is the spin field of vacuum electromagnetism, and $B^{(0)}$ is the scalar amplitude of the magnetic flux density of the beam. The conjugate product $\boldsymbol{B}^{(1)} \times \boldsymbol{B}^{(2)}$ is the basis of magneto-optical phenomena [7], of which there are several well known examples [13], and so $\boldsymbol{B}^{(3)}$ is an experimental observable. It magnetizes material matter, for example a plasma of electrons set up in helium by microwave pulses [14], and the magnetization, $\boldsymbol{M}^{(3)}$, set up by $\boldsymbol{B}^{(3)}$ is proportional to the *square root* of the beam power density (I_0), or intensity, in W m^{-2}. The required experimental conditions for the unequivocal isolation of the characteristic $I_0^{1/2}$ profile of $\boldsymbol{B}^{(3)}$ have been determined precisely [6] by solving the relativistic Hamilton-Jacobi equation of one electron (e) in the classical electromagnetic field, represented by the four-potential A_μ.

The various consequences of $\boldsymbol{B}^{(3)}$ have been worked into several branches of contemporary electromagnetic field theory [5—12], but in this Letter, its effect is explored on electroweak theory, which unifies electromagnetism with the weak field [15]. It is shown in Sec. 6.2 that the existence of the observable $\boldsymbol{B}^{(3)}$ in electromagnetism means that the gauge group symmetry must be enlarged to *O(3)*. In Sec. 6.3 it is shown that this means that the massive bosons of *GWS* acquire an additional physical dimension, the (3) dimension, and their concomitant fields are no longer purely transverse. In other words, the observable $\boldsymbol{B}^{(3)}$ of the electromagnetic sector is made up of conjugate products of intermediate vector boson field components. The latter are therefore also experimental observables. In Sec. 6.4, finally, it is shown that the observable $\boldsymbol{B}^{(3)}$ in electromagnetism does not affect the ability of *GWS* to predict the correct intermediate vector boson masses.

6.2 The O(3) Gauge Group of Vacuum Electromagnetism

The defining Lie algebra (1) is that of the non-Abelian group of rotations, *O(3)*, in three dimensional space [6,15]. Since $\boldsymbol{B}^{(3)}$ is a physical observable, the gauge group of vacuum electromagnetism is also *O(3)*, and not the *O(2)* of the conventional view [15]. There is a physical magnetic flux density, $\boldsymbol{B}^{(3)}$, orthogonal to the plane of definition of the plane waves $\boldsymbol{B}^{(1)} = \boldsymbol{B}^{(2)*}$. The photon, therefore, can no longer be regarded as a particle without mass, because special relativity [15] shows that such a particle can have only two (transverse) degrees of polarization. The Wigner little group [16] for the photon as particle also becomes *O(3)*, and not the obscure *E(2)*, the group of rotations and translations in a plane. The inference of photon mass leads in turn to the replacement [6] of the d'Alembert with the Proca equation, which leads to the replacement of $B^{(0)}$ in Eqs. (2.6.1) by the very slowly exponentially decaying $B^{(0)} \exp(-\xi Z)$ where ξ is the photon rest wavenumber [6] and Z is distance along the direction of propagation of radiation in vacuo. The range of electromagnetism is therefore not infinite, as discussed by Vigier [17].

The inference of an *O(3)* gauge group leads, furthermore, to a generalization of the vacuum Maxwell equations [6] to take into account the existence of a physical third axis (3) in free space. The usual plane wave relations are supplemented by an equation formally linking $\boldsymbol{B}^{(3)}$ and the imaginary and unphysical electric field strength $i\boldsymbol{E}^{(3)}$ in free space

$$\nabla \times \left(i\boldsymbol{E}^{(3)} \right) = -\frac{\partial \boldsymbol{B}^{(3)}}{\partial t} = \boldsymbol{0} . \tag{2.6.2}$$

The defining Lie algebra for $i\boldsymbol{E}^{(3)}$ links it to the ordinary plane wave $\boldsymbol{E}^{(1)} = \boldsymbol{E}^{(2)*}$ through the cyclically symmetric,

$$\boldsymbol{E}^{(1)} \times \boldsymbol{E}^{(2)} = -E^{(0)}\left(i\boldsymbol{E}^{(3)}\right)^*, \qquad \boldsymbol{E}^{(2)} \times \left(i\boldsymbol{E}^{(3)}\right) = -E^{(0)}\boldsymbol{E}^{(1)*},$$
$$\left(i\boldsymbol{E}^{(3)}\right) \times \boldsymbol{E}^{(1)} = -E^{(0)}\boldsymbol{E}^{(2)*}. \tag{2.6.3}$$

In contrast to the Lie algebra of magnetic fields, Eqs. (2.6.1), the conjugate product of polar vectors $\boldsymbol{E}^{(1)} \times \boldsymbol{E}^{(2)}$ cannot form a real polar vector, only the axial $\boldsymbol{B}^{(3)}$ [5],

$$\boldsymbol{E}^{(1)} \times \boldsymbol{E}^{(2)} = c^2\boldsymbol{B}^{(1)} \times \boldsymbol{B}^{(2)} = ic^2B^{(0)}\boldsymbol{B}^{(3)*}, \tag{2.6.4}$$

so $i\boldsymbol{E}^{(3)}$ in Eqs. (2.6.4) is mathematically an axial vector. It is therefore unphysical because a physical electric field is a polar vector, and indeed there are no known effects of the putative physical $\boldsymbol{E}^{(3)}$. In contrast, $\boldsymbol{B}^{(3)}$ is a real, axial vector, i.e., has the necessary symmetry for a physical magnetic field. The latter is therefore an experimental observable, *the first classical vacuum field to be inferred since Maxwell.* It is the spin field (i.e., phase free magnetic field) fundamentally responsible for *all* magneto-optic phenomena. For example, the well known inverse Faraday effect [13,14] can be understood [18] at visible frequencies in terms of the conjugate product, which is now understood to be the product $iB^{(0)}\boldsymbol{B}^{(3)*}$, and this is now recognized to be the second order component. There is also a first order component of the inverse Faraday effect due to $\boldsymbol{B}^{(3)}$ itself [6]. This dominates at microwave frequencies with an $I_0^{1/2}$ dependence under well defined conditions [19].

Since unified field theory such as *GWS* is based conventionally on the assumption that the electromagnetic gauge group is *O(2) (= U(1))*, it has to be re-examined as follows in the light of $\boldsymbol{B}^{(3)}$.

6.3 The Effect of $\boldsymbol{B}^{(3)}$ on GWS

The enlargement of the *O(2)* sector of *GWS* to *O(3)* must occur in such a way that it maintains the ability of *GWS* to predict the correct intermediate boson masses of the well known CERN experiment [15]. Obviously, the observed masses cannot change with the belated realization that $\boldsymbol{B}^{(3)}$ exists in the vacuum, and so $\boldsymbol{B}^{(3)}$ cannot affect the boson masses. In *GWS*, the potential four-vector A_μ is expressed in terms of the massive bosons $W_{3\mu}$ and X_μ which are components of the electromagnetic field,

$$A_\mu = W_\mu^3 \sin\theta_W + X_\mu \cos\theta_w \, . \tag{2.6.5}$$

Here θ_w is the Weinberg angle, which is fixed experimentally. So the extent to which W_μ^3 and X_μ can contribute to A_μ is also fixed experimentally. In an abstract isospin space [15], the physical part of W_μ^3 is the 3 component of this abstract space. X_μ on the other hand is an isospin scalar [15]. In the four dimensional space-time of special relativity, however, both W_μ^3 and X_μ are four-vectors, and can therefore be written in Minkowski notation as

$$W_\mu^3 := \left(\boldsymbol{W}^3, iW^{3(0)}\right), \qquad X_\mu := \left(\boldsymbol{X}, iX^{(0)}\right). \tag{2.6.6}$$

The space parts can be expressed in the circular basis (1), (2), (3) giving

$$\begin{aligned} \boldsymbol{A}^{(1)} &= \boldsymbol{W}^{3(1)} \sin\theta_w + \boldsymbol{X}^{(1)} \cos\theta_w \, , \\ \boldsymbol{A}^{(2)} &= \boldsymbol{W}^{3(2)} \sin\theta_w + \boldsymbol{X}^{(2)} \cos\theta_w \, , \end{aligned} \tag{2.6.7}$$

for $\boldsymbol{A}^{(1)}$ and its complex conjugate $\boldsymbol{A}^{(2)}$. Therefore $\boldsymbol{B}^{(3)}$ can be expressed in terms of transverse components of the massive bosons W_μ^3 and X_μ by

$$\boldsymbol{B}^{(3)} = -i\frac{\kappa^2}{B^{(0)}}\boldsymbol{A}^{(1)} \times \boldsymbol{A}^{(2)}, \tag{2.6.8}$$

where κ is the wavenumber of the electromagnetic beam; from this we infer that the bosons W^3_μ and X_μ themselves have longitudinal components which define Lie algebras akin to Eqs. (2.6.1) and (2.6.3). Since W^3_μ and X_μ are parts of A_μ in *GWS*, they are plane waves, e.g.,

$$\boldsymbol{W}^{3(1)} = \frac{W^{3(0)}}{\sqrt{2}}(i\boldsymbol{i} + \boldsymbol{j})\, e^{i\phi},$$
$$\boldsymbol{W}^{3(2)} = \frac{W^{3(0)}}{\sqrt{2}}(-i\boldsymbol{i} + \boldsymbol{j})\, e^{-i\phi}, \tag{2.6.9}$$

and so

$$\boldsymbol{W}^{3(1)} \times \boldsymbol{W}^{3(2)} = iW^{3(0)}\boldsymbol{W}^{3(3)*}$$
$$= -W_3^{(0)}\left(i\boldsymbol{W}^{3(0)}\right)^*, \qquad \text{et. cyclicum} \tag{2.6.10}$$

is a Lie algebra akin to (3), $\boldsymbol{W}^{3(1)}$ and $\boldsymbol{W}^{3(2)}$ being polar vectors, parts of $\boldsymbol{A}^{(1)}$ and $\boldsymbol{A}^{(2)}$ respectively. Similarly,

$$\boldsymbol{X}^{(1)} \times \boldsymbol{X}^{(2)} = -X^{(0)}\,(i\boldsymbol{X}^{(3)})^* \tag{2.6.11}$$

is a Lie algebra. Thus, both W^3_μ and X_μ are described by *O(3)* gauge geometry. We have therefore succeeded in enlarging the gauge geometry of GWS to include $\boldsymbol{B}^{(3)}$ self-consistently.

6.4 Boson Masses and $\boldsymbol{B}^{(3)}$

With the advent of $\boldsymbol{B}^{(3)}$ in the electromagnetic field, the W_μ^3 and X_μ bosons acquire three states of circular polarization, (1), (2) and (3). The extra state of polarization does not affect the mass of the boson. For example, we have

$$W_\mu^3 W_\mu^3 = \boldsymbol{W}^{3(1)} \cdot \boldsymbol{W}^{3(1)} + \boldsymbol{W}^{3(2)} \cdot \boldsymbol{W}^{3(2)} + \boldsymbol{W}^{3(3)} \cdot \boldsymbol{W}^{3(3)} - W^{3(0)2} , \tag{2.6.12}$$

which contains the additional term $\boldsymbol{W}^{3(3)} \cdot \boldsymbol{W}^{3(3)} - W^{3(0)2}$, a part of the additional term $\boldsymbol{A}^{3(3)} \cdot \boldsymbol{A}^{3(3)} - A^{3(0)2}$ in electromagnetism. However, this term vanishes because $|\boldsymbol{A}^{(3)}| = A^{(0)}$. The mass of W_μ^3 appears as a premultiplier of $W_\mu^3 W_\mu^3$ in the appropriate Lagrangian [15], and from this we infer that the extra (3) polarization makes no difference to the mass of the boson concomitant with W_μ^3. The only way in which the mass could be affected were if the premultiplier were for some reason different for transverse and longitudinal terms. This does not seem very likely because mass is a scalar Lorentz invariant. Four-vector products such as $W_\mu^3 W_\mu^3$ and $X_\mu X_\mu$ are also Lorentz invariants.

In conventional *GWS* [15], the photon mass is modeled to zero, but the concept of spontaneous symmetry breaking of the vacuum is used within non-Abelian, abstract isospin space [15], to provide the intermediate vector bosons with mass. The advent of $\boldsymbol{B}^{(3)}$, however, means that the photon must also be massive in *GWS*. This is a direct result of the experimental observable $\boldsymbol{B}^{(3)}$ which was related in Sec. 6.3 to the vector bosons. The latter acquire in turn the polarization (3), which cannot exist in a massless photon. This implies that *GWS* (and grand unified theory such as *SU(5)*) must accommodate finite photon mass, for example as in the work of Huang

[1]. This illustrates how $\boldsymbol{B}^{(3)}$ has highly non-trivial repercussions throughout contemporary unified and grand unified field theory. In the electromagnetic sector, the Higgs mechanism is well known to be compatible with gauge invariance, and leads to finite photon mass through spontaneous symmetry breaking of the vacuum. The acquired photon mass is inevitably accompanied [15] by the acquisition of a third, physical polarization, manifest in $\boldsymbol{B}^{(3)}$. This type of result is, however, modeled out in *GWS* to force the result that photon mass is identically zero. With the advent of $\boldsymbol{B}^{(3)}$ such a procedure is invalidated and must be replaced by a mechanism which self-consistently accounts for photon mass.

Acknowledgments

It is a pleasure to acknowledge many interesting discussions at various formative stages of $\boldsymbol{B}^{(3)}$ theory, with, among others: Yildirim Aktas, the late Stanisław Kielich, Stuart Kurtz, Mikhail Novikov, Mark Silverman and Jean-Pierre Vigier.

References

[1] J. C. Huang, *J. Phys., G, Nucl. Phys.* **13**, 273 (1987).
[2] S. Weinberg, *Rev. Mod. Phys.* **52**, 515 (1980).
[3] A. Salam, *Rev. Mod. Phys.* **52**, 525 (1980).
[4] S. L. Glashow, *Rev. Mod. Phys.* **52**, 539 (1980).
[5] M. W. Evans, *Physica B* **182**, 227, 237 (1992); **183**, 103 (1993); **190**, 310 (1993).
[6] M. W. Evans and J.-P. Vigier, *The Enigmatic Photon, Volume 1, The Field* $\boldsymbol{B}^{(3)}$ (Kluwer, Dordrecht, 1994); *The Enigmatic Photon, Volume 2, Non-Abelian Electrodynamics* (Kluwer Academic, Dordrecht,1995).

[7] M. W. Evans, and S. Kielich, eds., *Modern Nonlinear Optics,* Vol. 85(2) of *Advances in Chemical Physics,* I. Prigogine and S. A. Rice, eds. (Wiley Interscience, New York, 1993/1994).
[8] M. W. Evans and A. A. Hasanein, *The Photomagneton in Quantum Field Theory* (World Scientific, Singapore, 1994).
[9] M. W. Evans in A. Garuccio and A. van der Merwe eds., *Waves and Particles in Light and Matter* (Plenum, London, 1994), de Broglie Centenary Volume.
[10] M. W. Evans, *Mod. Phys. Lett.* **7**, 1247 (1993); *Found. Phys. Lett.* **7**, 67 (1994).
[11] M. W. Evans, *Found. Phys. Lett.* and *Found. Phys.* in press, 1994 / 1995.
[12] M. W. Evans, *Found. Phys. Lett.* submitted for publication.
[13] reviewed by R. Zawodny, in Vol. 85(1) of Ref. 7.
[14] J. Deschamps, M. Fitaire, and M. Lagoutte, *Phys. Rev. Lett.* **25**, 1330 (1970); *Rev. Appl. Phys.* **7**, 155 (1972).
[15] L. H. Ryder, *Quantum Field Theory*, 2nd edn. (Cambridge University Press, Cambridge, 1987)
[16] E. P. Wigner, *Ann. Math.* **40**, 149 (1939).
[17] J.-P. Vigier, "Experimental Status of the Einstein-de Broglie Theory of Light." *Proc. I.S.Q.M. Workshop* (Tokyo, 1992).
[18] S. Woźniak, M. W. Evans and G. Wagnière, *Mol. Phys.* **75**, 81 (1992).
[19] M. W. Evans, *Found. Phys. Lett.* submitted for publication.

Paper 7

The Physical Meaning of $\boldsymbol{B}^{(3)}$

The physical meaning is discussed, using simple concepts, of the novel longitudinal field $\boldsymbol{B}^{(3)}$ (the Evans-Vigier field) of vacuum electromagnetism. In words without equations, it is explained why the physical $\boldsymbol{B}^{(3)}$ is not accompanied by a physical electric field. The source of $\boldsymbol{B}^{(3)}$; its mode of propagation; and its symmetry and energy characteristics are explained in physical terms rather than mathematical.

Key words: $\boldsymbol{B}^{(3)}$ field, physical meaning of.

7.1 Introduction

In recent months [1—10] it has become clear that the conventional view of vacuum electromagnetism is incomplete, because there exists in the vacuum the Evans-Vigier field, $\boldsymbol{B}^{(3)}$. This is a novel, classical, magnetic field in free space, the first to be inferred since Maxwell. As such, it has its quantum mechanical counterpart, the photomagneton [2]. The purpose of this short note is to describe the physical meaning of $\boldsymbol{B}^{(3)}$ without abstract mathematics, because $\boldsymbol{B}^{(3)}$ is a remarkable development, having been hidden in Maxwell's equations for well over a hundred years. In Sec. 7.2, its origin is described in the plane waves of vacuum electromagnetism. In Sec. 7.3, it is explained in physical terms why $\boldsymbol{B}^{(3)}$ propagates through the vacuum, and finally, in Sec. 7.4, its symmetry and energy characteristics are explained in physical terms as simply as words allow. A short discussion section explains how $\boldsymbol{B}^{(3)}$ can be isolated from the ever-present plane waves which are its source in the vacuum, and thereby observed experimentally.

7.2 The Origin of the Evans-Vigier Field

The magnetic component of the plane wave, $\boldsymbol{B}^{(1)}$, from Maxwell's equations is a complex quantity in general and so has a complex conjugate, $\boldsymbol{B}^{(2)}$, which is also a solution of Maxwell's equations. This can be thought of in terms of a complex, circular, representation of three dimensional space, a representation which is entirely equivalent to the usual real Cartesian. This picture is analogous to, but not the same as, a complex spinor representation used routinely for fermions such as the electron. The vectors $\boldsymbol{B}^{(1)}$ and $\boldsymbol{B}^{(2)}$ are therefore components of the complete vector field in this circular representation [11]. They are, however, only two components out of a possible three, because we are dealing with three dimensional space. The third component is the Evans-Vigier field, denoted as $\boldsymbol{B}^{(3)}$ because (3) is the third axis associated with (1) and (2). The component $\boldsymbol{B}^{(3)}$ is generated by the vector product of $\boldsymbol{B}^{(1)}$ and $\boldsymbol{B}^{(2)}$ in analogy with $\boldsymbol{i}\ x\ \boldsymbol{j} = \boldsymbol{k}$, et cyclicum, of the Cartesian representation, where **i**, **j**, and **k** are the usual Cartesian unit vectors along *X*, *Y* and *Z* respectively.

Very simply, therefore, the physical source of $\boldsymbol{B}^{(3)}$ is the cross product $\boldsymbol{B}^{(1)} \times \boldsymbol{B}^{(2)}$. This inference is cyclically symmetric [6—10], the physical source of $\boldsymbol{B}^{(1)}$ is $\boldsymbol{B}^{(2)} \times \boldsymbol{B}^{(3)}$ and that of $\boldsymbol{B}^{(2)}$ is $\boldsymbol{B}^{(3)} \times \boldsymbol{B}^{(1)}$. The three components $\boldsymbol{B}^{(1)}$, $\boldsymbol{B}^{(2)}$ and $\boldsymbol{B}^{(3)}$ are physical magnetic fields and all three are axial vectors. The cross product of two axial vectors is another axial vector, so that this view is self-consistent. The source of the Evans-Vigier field is therefore the plane wave components $\boldsymbol{B}^{(1)}$ and $\boldsymbol{B}^{(2)}$.

It can be shown [10] that $\boldsymbol{B}^{(3)}$ can be expressed consistently in a variety of ways. The source of a magnetic field in classical electrodynamics [12] is described by the Biot-Savart-Ampère (BSA) law, and, indeed, $\boldsymbol{B}^{(3)}$ can be written [10] in this form, i.e., as the vector cross product of a transverse momentum, $\boldsymbol{p}^{(1)}$, with an electric component, $\boldsymbol{E}^{(2)}$, of the plane wave. Another inference of classical electrodynamics is that a magnetic field is the curl of a vector potential, and, indeed, $\boldsymbol{B}^{(3)}$ can be written as $-\nabla^{(1)} \times \boldsymbol{A}^{(2)}$, where $\nabla^{(1)}$ is a well defined curl operator [10] and $\boldsymbol{A}^{(2)}$ a plane wave potential [10]. Furthermore, the *BSA* and *curl A* forms of $\boldsymbol{B}^{(3)}$ are equivalent to the various *double field* forms typified by $\boldsymbol{B}^{(1)} \times \boldsymbol{B}^{(2)}$. This analysis, developed elsewhere [10] shows that in the classical sense, $\boldsymbol{B}^{(1)} \times \boldsymbol{B}^{(2)}$ is indeed a source for a magnetic field. There are also various other forms of $\boldsymbol{B}^{(3)}$, tabulated in the literature [10], and its existence has been demonstrated from first principles using the Dirac equation [7] of relativistic quantum field theory, and the Hamilton-Jacobi equation [6] of relativistic classical field theory. Intuitively, the complicated language and mathematical analysis behind these demonstrations can be reduced to a consideration of the helical motion of the tip of a magnetic or electric field propagating in free space. The field is $\hat{C}$ negative, like charge, and so this motion is intuitively analogous to that of a current through a solenoid, producing a magnetic field. It is well known [13] in the classical theory of fields that a circularly polarized electromagnetic wave drives an electron in a circle, so the field itself is in a sense, charged, (i.e., $\hat{C}$ negative, where $\hat{C}$ is the charge conjugation operator) otherwise there would be no effect on the electron. The photon, on the other hand, is considered to be an uncharged quantum of energy. Therefore the photon and field must always be considered concomitantly.

7.3 Propagation Through the Vacuum of the Evans-Vigier Field

The cross product $\boldsymbol{B}^{(1)} \times \boldsymbol{B}^{(2)}$ removes the electromagnetic phase, ϕ [1—10], so that $\boldsymbol{B}^{(3)}$ has been described as a static magnetic field. More accurately, it is a phase free magnetic field, and propagates through the vacuum with its source, the plane wave components $\boldsymbol{B}^{(1)}$ and $\boldsymbol{B}^{(2)}$. Therefore, a pulse of light carries with it the source of a $\boldsymbol{B}^{(3)}$ field through the vacuum, and so does a continuous beam. The pulse is detectible experimentally because it has travelling intensity, or power density (W m^{-2}), and the cross product $\boldsymbol{B}^{(1)} \times \boldsymbol{B}^{(2)}$ is physically interpretable as being proportional to the antisymmetric part of the light intensity tensor [3,14] itself. This is always non-zero, and so is $\boldsymbol{B}^{(3)}$. Therefore $\boldsymbol{B}^{(3)}$ propagates because the intensity of light propagates. The intensity of light, even in the vacuum, is a tensor [3,14,15], this being an early inference by Placzek in the theory of non-linear optics [14]. (Intensity is quadratic in the electric field of the plane wave.) If light intensity did not have an antisymmetric, mathematically imaginary, tensor component, there would be no inverse Faraday effect [16], in which light magnetizes material matter, a process which is free of the electromagnetic phase as first inferred by Pershan [16] using general arguments. The inverse Faraday effect has been observed experimentally in glasses and liquids [17] and in an electron plasma [18] using respectively visible and microwave radiation. Since $\boldsymbol{B}^{(1)} \times \boldsymbol{B}^{(2)}$ is the source of $\boldsymbol{B}^{(3)}$, the inverse Faraday effect is due to $\boldsymbol{B}^{(3)}$, and it is concluded that $\boldsymbol{B}^{(3)}$ is the fundamental field of magneto-optics [1—10]. Intuitively, it is expected that magnetization be due to a magnetic field, and magnetization by light is due to the magnetic field $\boldsymbol{B}^{(3)}$, a consistent physical result. We conclude that the antisymmetric part of light intensity manifests itself physically as a magnetic field, the Evans-Vigier field.

7.4 Considerations of Symmetry and Energy Conservation

The cross product of two axial vectors is another axial vector [19], and the cross product of two polar vectors is also an axial vector. Thus, both $\boldsymbol{B}^{(1)} \times \boldsymbol{B}^{(2)}$ and the cross product of electric plane wave components, $\boldsymbol{E}^{(1)} \times \boldsymbol{E}^{(2)}$, are proportional to $\boldsymbol{B}^{(3)}$ [6]. It can also be shown that the cross product of vector potentials, $\boldsymbol{A}^{(1)} \times \boldsymbol{A}^{(2)}$, is similarly proportional to $\boldsymbol{B}^{(3)}$, a result which leads to a self-consistent representation of $\boldsymbol{B}^{(3)}$ using *O(3)* gauge theory rather than the conventional *O(2)* = *U(1)* symmetry group [19] for the electromagnetic sector in contemporary field theory. The various technical ramifications of the *O(3)* gauge group are developed elsewhere [6]. These include the important inference that the photon as particle must have mass, because it is three dimensional in nature, not two, as in the conventional *O(2)* gauge group. Here we are concerned with a simpler inference of symmetry, that $\boldsymbol{B}^{(3)}$ cannot be accompanied by a real $\boldsymbol{E}^{(3)}$, an inference which follows from the fact that a real polar vector cannot be formed from the conjugate products $\boldsymbol{B}^{(1)} \times \boldsymbol{B}^{(2)}$, $\boldsymbol{E}^{(1)} \times \boldsymbol{E}^{(2)}$, or $\boldsymbol{A}^{(1)} \times \boldsymbol{A}^{(2)}$. It can be shown [6] that $\boldsymbol{B}^{(3)}$ is accompanied, formally, by a pure imaginary $-i\boldsymbol{E}^{(3)}$, which being imaginary and first order, is not a physical field. Consistently, no first order experimental effect of a putative $\boldsymbol{E}^{(3)}$ has been reported. This is again an intuitively comfortable result because we do not expect a solenoid to produce an electric field in its axis, only a magnetic field. In this intuitive view, light is, loosely speaking, an optical solenoid producing $\boldsymbol{B}^{(3)}$, an axial vector about its axis of propagation. We have therefore referred to $\boldsymbol{B}^{(3)}$ as the spin field [6—10] to distinguish it from the wave field components $\boldsymbol{B}^{(1)}$ and $\boldsymbol{B}^{(2)}$.

When considering the effect of $\boldsymbol{B}^{(3)}$ on electromagnetic energy density, however, it is necessary to consider vector magnitudes. In the circular representation (1), (2), (3), this means that we must consider dot products $\boldsymbol{B}^{(i)} \cdot \boldsymbol{B}^{(i)*}$, where * denotes complex conjugate and where i runs from 1 to 3, not from 1 to 2 as in the conventional view [12]. (Recall that in the conventional view there are only wave components, $\boldsymbol{B}^{(1)}$ and $\boldsymbol{B}^{(2)}$, and these exist in a flat, two dimensional, world.) The quantum of light energy, $\hbar\omega$, the dictionary photon [19] must then be expressed in terms of the sum $\boldsymbol{B}^{(1)} \cdot \boldsymbol{B}^{(1)*} + \boldsymbol{B}^{(2)} \cdot \boldsymbol{B}^{(2)*} + \boldsymbol{B}^{(3)} \cdot \boldsymbol{B}^{(3)*}$. This does not change the Planck constant, however, because the same quantum of energy, $\hbar\omega$, is merely redistributed among (or thought of in terms of) three vector components, the overall magnitude remaining the same [10]. (One photon remains one photon, and can be detected experimentally as such. We infer that it is concomitant with three field components rather than two as thought conventionally.) Similarly, the quantum of light energy can be thought of in terms of the sum of three electric field components $\boldsymbol{E}^{(1)} \cdot \boldsymbol{E}^{(1)*} + \boldsymbol{E}^{(2)} \cdot \boldsymbol{E}^{(2)*} + i\boldsymbol{E}^{(3)} \cdot \left(i\boldsymbol{E}^{(3)}\right)^*$, where, now, $-i\boldsymbol{E}^{(3)}$ is multiplied by its own conjugate and becomes real and therefore physical. In the vacuum, the sum of magnetic field components is proportional to the sum of electric field components, both being proportional to $\hbar\omega$. The quantum of energy is again redistributed among three concomitant field components rather than two, its magnitude, and that of Planck's constant h, being unchanged. This energy analysis and others like it [6] illustrates the need for $-i\boldsymbol{E}^{(3)}$ as well as $\boldsymbol{B}^{(3)}$. Since $\hbar\omega$ is unchanged it is concluded that $\boldsymbol{B}^{(3)}$ does not affect the fundamentals of the old quantum theory, e.g. the Einstein theory of absorption and spontaneous emission, and the light quantum hypothesis itself. Thus spectra remain frequency dependent and discrete, $\boldsymbol{B}^{(3)}$ is a magnetizing field that is phase free, and frequency independent. The theory of the inverse Faraday effect is the same in structure at microwave or visible frequencies far from optical resonance. At or near resonance, Woźniak *et al.* [20] have shown that useful additional frequency dependent features occur but the optical property fundamentally responsible for magnetization by light remains the conjugate product, which is now understood [1—10] to be $iB^{(0)}\boldsymbol{B}^{(3)*}$, where $B^{(0)}$ is the field amplitude. It should be clearly understood that although the conjugate product is imaginary, $\boldsymbol{B}^{(3)}$ itself is real, and physical. This is, at the root, a consequence of space geometry itself [6].

7.5 Discussion

It has just been inferred that $\boldsymbol{B}^{(3)}$ is real and physical, so must produce real and physical effects which are observable experimentally. Conversely, $-i\boldsymbol{E}^{(3)}$ is unphysical, and cannot produce observable effects at first order. In order to separate out, or isolate, $\boldsymbol{B}^{(3)}$ experimentally, and thus to prove its existence, it is necessary to demonstrate its characteristic square root power density ($I_0^{1/2}$) dependence [6]. The classical, but relativistic, theory of the orbital angular momentum of the electron in the electromagnetic field is adequate [6] to show that the $I_0^{1/2}$ dependence can be expected to dominate experimentally using microwave pulses of sufficient intensity, or power density. An increase in power density of about two orders of magnitude over that used by Deschamps *et al.* [18] should be sufficient. The $I_0^{1/2}$ dependence of the magnetization can be due only to the vacuum Evans-Vigier field, because first order magnetization effects due to the plane waves $\boldsymbol{B}^{(1)}$ and $\boldsymbol{B}^{(2)}$ disappear on average. In general, the interaction of $\boldsymbol{B}^{(3)}$ with one electron is relativistic in nature, at visible frequencies the magnetizing effect of light is dominated [6] by I_0 acting at first order, at microwave frequencies, with sufficient power density by $I_0^{1/2}$. This result explains why the measurements on the inverse Faraday effect to date [17,18] have shown an I_0 dependence of the magnetization. The $I_0^{1/2}$ dependence has also emerged from an interesting analysis by Chiang [21] of the influence of ion motion in the inverse Faraday effect. His figure one shows the expected linear dependence of magnetization on $\boldsymbol{B}^{(3)}$, but Chiang did not make the key inference that $\boldsymbol{B}^{(3)}$ is generated [6] from $\boldsymbol{B}^{(1)} \times \boldsymbol{B}^{(2)}/(iB^{(0)})$ in free space. Chiang's analysis however agrees qualitatively with that given in this discussion, as the power density increases, the quadratic dependence of plasma magnetization on I_0 *evolves into an* $I_0^{1/2}$ *dependence*, which eventually saturates, because the maximum orbital angular momentum that the photon can transfer to the electron in a perfectly elastic collision is $\hbar$ [10]. We have also reached this conclusion independently [10], and physically, it means that the angular momentum of the photon, $\hbar$, has been completely transferred to the electron using enormous beam power densities. An experimental investigation of these effects is necessary, because data to date have been confined to the quadratic region, (I_0 dependence of the magnetization).

Acknowledgments

It is a pleasure to acknowledge many interesting discussions with several colleagues, among whom are: Gareth J. Evans, the late Stanislaw Kielich, Mikhail A. Novikov, Mark P. Silverman, Jean-Pierre Vigier, and Boris Yu. Zel'dovich.

References

[1] M. W. Evans, *Physica B* **182**, 227, 237 (1992).
[2] M. W. Evans, *Physica B* **183**, 103 (1993); **190**, 310 (1993).
[3] M. W. Evans, *Mod. Phys. Lett.* **7**, 1247 (1993).
[4] M. W. Evans, *Found. Phys. Lett.* **7**, 67 (1994).
[5] M. W. Evans, *The Photon's Magnetic Field.* (World Scientific, Singapore, 1992).
[6] M. W. Evans and J.-P. Vigier, *The Enigmatic Photon, Volume 1: The Field* $\boldsymbol{B}^{(3)}$ (Kluwer, Dordrecht, 1994).
[7] ibid., *The Enigmatic Photon, Volume 2: Non-Abelian Electrodynamics* (Kluwer Academic, Dordrecht, 1995).
[8] M. W. Evans and A. A. Hasanein, *The Photomagneton and Quantum Field Theory*, Vol. 1 of *Quantum Chemistry.* (World Scientific, Singapore, 1994), a collection of articles on the $\boldsymbol{B}^{(3)}$ field and related topics.
[9] M. W. Evans, and S. Kielich, eds., *Modern Nonlinear Optics,* Vol. 85(2) of *Advances in Chemical Physics,* I. Prigogine and S. A. Rice, eds., (Wiley Interscience, New York, 1993).

[10] M. W. Evans, *Found. Phys. Lett.* and *Found. Phys.* in press. Also, M. W. Evans in A. van der Merwe and A. Garuccio eds., *Waves and Particles in Light and Matter.* (Plenum, New York, 1994), a collection of papers on the occasion of the centennial of the birth of Prince Louis Victor de Broglie.

[11] R. Zawodny, in Vol. 85(1) of Ref. 9.

[12] J. D. Jackson, *Classical Electrodynamics* (Wiley, New York, 1962).

[13] L. D. Landau and E. M. Lifshitz, *The Classical Theory of Fields*, 4th edn. (Pergamon, Oxford, 1974).

[14] S. Kielich, in M. Davies ed. *Dielectric and Related Molecular Processes*, Vol. 1 (Chemical Society, London, 1972).

[15] G. H. Wagnière, *Linear and Nonlinear Optical Properties of Molecules.* (VCH, Basel, 1993).

[16] P. S. Pershan, *Phys. Rev.* **130**, 919 (1963).

[17] J. P. van der Ziel, P. S. Pershan, and L. D. Malmstrom, *Phys. Rev.* **143**, 574 (1966).

[18] J. Deschamps, M. Fitaire, and M. Lagoutte, *Phys. Rev. Lett.* **25**, 1330 (1970); *Rev. Appl. Phys.* **7**, 155 (1972).

[19] L. D. Barron, *Molecular Light Scattering and Optical Activity* (Cambridge University Press, Cambridge, 1982).

[20] S. Woźniak, M. W. Evans, and G. Wagnière, *Mol. Phys* **75**, 81, 99 (1992).

[21] A. C.-L. Chiang, *Phys. Fluids* **24**, 369 (1981).

Paper 8

Relativistic Magneto-Optics and the Evans-Vigier Field

Relativistic effects in magneto-optics are discussed in order to isolate experimentally the newly inferred Evans-Vigier field of vacuum electromagnetism. The discussion is reduced to its simplest form by considering the interaction of one photon with one electron over the complete range of photon energy and momenta transfer. It is shown that there are three regions: a) quadratic, b) linear and c) region of saturation; into which the characteristics of the interaction process can be divided. In region b), the angular momentum imparted to the electron by the photon is linearly dependent on the Evans-Vigier field $\boldsymbol{B}^{(3)}$, and this region is accessible experimentally with microwave pulses of sufficient power density. In magneto-optics with visible lasers, only region a) is accessible, and this is shown to be the non-relativistic limit.

Key words: relativistic magneto-optics, Evans-Vigier field.

8.1 Introduction

Magneto-optic effects received great impetus from the discovery of the laser, but their existence was inferred earlier [1]. They have recently been reviewed comprehensively by Zawodny [2], using the accepted semi-classical approach [3]. It is shown in this Letter that this is the non-relativistic limit, suitable for application in what we infer here to be the *quadratic region* (a). This region is characterized by a relatively low beam power density (I_0 in W m^{-2}) and high beam angular frequency (ω in rad sec^{-1}), conditions which obtain in the visible for all but the most powerful pulses. Under these conditions, the semi-classical approach (which is based on the non-relativistic Schrödinger equation [4]) shows that magnetization by light (the inverse Faraday effect [5,6]) is proportional to I_0. This has been verified experimentally in glasses and liquids [6] and in an electron plasma [7]. For many years, therefore, the theory has been thought of as complete, and magnetization by circularly polarized light has been ascribed to the well known conjugate product [8], the antisymmetric component of the light intensity tensor.

In this Letter it is shown that the above represents only the non-relativistic limit of magneto-optics. It is shown by considering the collision of a photon with an electron that as the beam power density is increased, and the beam angular frequency decreased to the microwave range, the magnetization of the inverse Faraday effect becomes proportional to $I_0^{1/2}$, which means that it is proportional at first order to a magnetic field $\boldsymbol{B}^{(3)}$, the newly inferred [9—14] Evans-Vigier field of vacuum electromagnetism. This is labeled as *region b)* or *linear region*, and can be described only with a correctly relativistic theory. As the beam power density is increased further for a given angular frequency, the curve of magnetization versus I_0 saturates, and we enter the *region of saturation*, region c). In this limit the magnetization is constant for all I_0. This is a purely relativistic phenomenon, with no non-relativistic meaning, and can be understood simply because the maximum angular momentum that the photon (at any beam frequency) can transfer to the electron is $\hbar$, the Dirac constant. In a perfectly elastic transfer of angular momentum, conservation demands that the angular momentum, $\hbar$, of the photon be annihilated and given up completely to the electron. This process, although allowed theoretically, requires enormous beam power densities.

In Sec. 8.2, the classical, but relativistic, Hamilton-Jacobi equation of one electron (e) in the electromagnetic field, represented by the potential four-vector A_μ, is used to illustrate the existence of regions a) and b), and to infer that region a) is the non-relativistic limit. Sec. 8.3 describes the development of region c), by quantizing the field into energy quanta, photons. Thereafter, Sec. 8.4 uses simple Compton effect theory to illustrate the existence of regions a), b) and c) in energy and linear momentum transfer from photon to electron.

8.2 Classical, Relativistic, Magneto-Optics

The correctly relativistic, but classical, theoretical basis for magneto-optics can be developed in a relatively simple way by using the Hamilton-Jacobi equation [15]. The trajectory of one electron in the electromagnetic field can be shown [15] to be governed by a classical, orbital, angular momentum,

$$\boldsymbol{J}^{(3)} = \frac{e^2c^2}{\omega^2}\left(\frac{B^{(0)}}{\left(m_0^2\omega^2 + e^2B^{(0)2}\right)^{1/2}}\right)\boldsymbol{B}^{(3)} . \tag{2.8.1}$$

Here e/m_0 is the charge to mass ratio of the electron, and $B^{(0)}$ the magnetic flux density amplitude of the beam. The magnetization induced by the beam is therefore due entirely to the Evans-Vigier field $\boldsymbol{B}^{(3)}$

$$\boldsymbol{M}^{(3)} = -\frac{e}{2m_0}\boldsymbol{J}^{(3)}. \tag{2.8.2}$$

In the condition,

$$\omega = \frac{e}{m_0} B^{(0)}, \tag{2.8.3}$$

this result becomes a sum of two terms [15],

$$\boldsymbol{M}^{(3)} = \frac{1}{2\sqrt{2}} \left(\chi' + B^{(0)} \beta'' \right) \boldsymbol{B}^{(3)}, \tag{2.8.4}$$

where

$$\chi' := -\frac{e^2 c^2}{2 m_0 \omega^2} \tag{2.8.5}$$

is the one electron susceptibility and where

$$\beta'' := -\frac{e^3 c^2}{2 m_0^2 \omega^3} \tag{2.8.6}$$

is the one electron hyperpolarizability. The relativistic factor of the Hamilton-Jacobi equation can be expressed in terms of the momentum magnitude $p = eA^{(0)}$, where $A^{(0)}$ is the amplitude of the vector potential of the beam,

$$\gamma := \frac{c}{\omega} \left(m_0^2 \omega^2 + e^2 B^{(0)2} \right)^{1/2} = \frac{1}{c} \left(m_0^2 c^4 + c^2 p^2 \right)^{1/2}. \tag{2.8.7}$$

In the non-relativistic limit, the speed of light is much greater than the speed, v, imparted to the electron (Sec. 8.4) by a collision with the photon. This means that

$$m_0 c \gg p\,, \tag{2.8.8}$$

a limit which corresponds to

$$\omega \gg \frac{e}{m_0} B^{(0)}\,, \tag{2.8.9}$$

i.e., the angular frequency of the beam is much greater than $eB^{(0)}/m_0$. For all but the most enormous laser power densities, this is always true in the visible, whereupon the expression for the magnetization in the one electron inverse Faraday effect becomes

$$\boldsymbol{M}^{(3)} \underset{c \gg v}{\rightarrow} \frac{\beta''}{2} B^{(0)} \boldsymbol{B}^{(3)}\,, \tag{2.8.10}$$

which is quadratic in $B^{(0)}$. Our quadratic region a) is therefore defined as the non-relativistic region of the interaction of one electron with the classical electromagnetic field. The result a) becomes recognizable in the conventional semi-classical theory [8] when the one electron hyperpolarizability is replaced by a semi-classical atomic or molecular hyperpolarizability, usually calculated [8] from a perturbation theory, using a quantum approach for the atom or molecule and a classical view of the field [4]. The only conceptual difference is that the atomic or molecular property tensor contains resonance features, and the free electron equivalent does not.

The classical version, $\boldsymbol{B}^{(3)}$, of the Evans-Vigier field therefore governs the inverse Faraday effect in atoms and molecules as well as in an electron plasma.

In the usual, non-relativistic, theory [8], however, the term linear in $\boldsymbol{B}^{(3)}$ is missing completely. This term becomes *dominant,* however, in the condition,

$$\omega \ll \frac{e}{m_0} B^{(0)}, \tag{2.8.11}$$

which obtains with microwave pulses [17,15] of sufficient power density. Under condition (2.8.11) we expect $\boldsymbol{M}^{(3)}$ to be proportional to $I_0^{1/2}$, and not to I_0 as in the visible. This expectation has yet to be verified experimentally, but it is nevertheless based on first principles. When $\boldsymbol{M}^{(3)}$ is linearly proportional to $\boldsymbol{B}^{(3)}$, we enter the linear region b). We cannot enter this region if our theory is non-relativistic, i.e., based on perturbation theory applied to the non-relativistic Schrödinger equation. The latter represents the usual semi-classical theory of the inverse Faraday effect. A more plausible semi-classical approach is one based on perturbation theory applied to the relativistic Dirac equation for free electrons, atoms and molecules. If done properly, this should correctly quantize our classical result (2.8.1), and extend it to atoms and molecules as well as the single free electron for which it is valid.

8.3 The Region of Saturation, c)

As the power density of the electromagnetic beam is increased in region b), it might be expected that $\boldsymbol{M}^{(3)}$ will simply increase indefinitely with $\boldsymbol{B}^{(3)}$. Special relativity shows that this is not the case, however, because when the electromagnetic field is quantized, the angular momentum of the photon is constant, $\hbar$ [15]. The law of conservation of angular momentum asserts that in a photon electron collision, the angular momentum transferred to the electron from the photon cannot exceed $\hbar$. Similarly, the energy and linear momentum transferred cannot exceed $\hbar\omega$ and $\hbar\kappa$ respectively, where

$\kappa = \omega/c$. Anticipating the quantized field theory, the maximum $\boldsymbol{J}^{(3)}$ from Eq. (2.8.1) is $\hbar \boldsymbol{e}^{(3)}$, where $\boldsymbol{e}^{(3)}$ is a unit vector [12],

$$\boldsymbol{J}^{(3)}_{\max} \rightarrow \hbar \boldsymbol{e}^{(3)}, \tag{2.8.12}$$

and the theory enters region c), the region of saturation. In region c), the magnetization in the inverse Faraday effect no longer depends on I_0. The physical meaning of this is that in the photon-electron collision, the $\hbar$ of the photon has been given up entirely to the electron in a perfectly elastic transfer of angular momentum.

Equation (2.8.12) can be obtained from Eq. (2.8.1) by using the charge quantization condition [15],

$$\hbar\kappa = eA^{(0)}, \tag{2.8.13}$$

in Eq. (2.8.1). In Eq. (2.8.13), the classical momentum magnitude $eA^{(0)}$ imparted to the electron by the field is identified with the quantized photon momentum, $\hbar\kappa$. Equation (2.8.1) becomes

$$\boldsymbol{J}^{(3)} = \frac{\hbar\kappa}{\left(m_0^2c^2 + \hbar^2\kappa^2\right)^{1/2}} \frac{ec^2}{\omega^2}\boldsymbol{B}^{(3)} . \tag{2.8.14}$$

This result clearly identifies the Evans-Vigier field $\boldsymbol{B}^{(3)}$ as solely responsible for the inverse Faraday effect. In the limit $\hbar\kappa \gg m_0 c$, which corresponds with $eB^{(0)} \gg m_0\omega$, Eq. (2.8.14) becomes

$$\boldsymbol{J}^{(3)} \rightarrow \frac{e}{\kappa^2}\boldsymbol{B}^{(3)}, \tag{2.8.15}$$

and using $B^{(0)} = \kappa A^{(0)}$

$$J^{(3)} \rightarrow \frac{eA^{(0)}}{\kappa} e^{(3)} = \hbar e^{(3)} . \tag{2.8.16}$$

This identifies the transition from region b) (Eq. (2.8.15)) to region c) (Eq. (2.8.16)). Note that this transition comes about through Eq. (2.8.13), which means that the maximum magnitude of the linear momentum imparted to the electron by the field is that of the photon. The law of conservation of momentum shows that this occurs in a perfectly elastic transfer of linear momentum $\hbar\kappa$ from the photon to the electron. The physical meaning of this is developed in the next section.

8.4 Regions A), B) and C) in Compton Theory

In this section, energy and linear momentum transfer in the photon-electron collision is analyzed with the simplest type of Compton theory. The key feature of the Compton effect is that the collision changes the frequency of the photon from ω_i to ω_f, providing early evidence for the light quantum hypothesis. Accordingly, the conservation of energy demands that [16]

$$\Delta En = \hbar \left(\omega_i - \omega_f \right) = \left(p^2 c^2 + m_0^2 c^4 \right)^{1/2} - m_0 c^2 , \tag{2.8.17}$$

where $\boldsymbol{p}$ is the linear momentum given to the electron (initially at rest) by the photon. Quantities on the left hand side of Eq. (2.8.17) refer to the photon, and those on the right hand side to the electron, whose rest energy is $m_0 c^2$. Similarly, conservation of linear momentum demands that

$$\hbar \left(\kappa_i - \kappa_f \right) = p . \tag{2.8.18}$$

If we consider the limit $\hbar\omega_i \gg m_0 c^2$, i.e., the collision of a very energetic photon with an initially stationary electron; and if the photon gives up its energy entirely to the electron, then $\omega_f = 0$. The translational kinetic energy acquired by the electron is such that $pc \gg m_0 c^2$, and so

$$\hbar\omega_i \sim pc\,. \tag{2.8.19}$$

The same result is obtained from Eq. (2.8.18) by assuming that $\kappa_f = 0$. i.e that the linear momentum of the photon is zero after collision and has been transferred elastically to the electron. (In an elastic collision, kinetic energy and linear momentum are both conserved.) In this limit,

$$\hbar\kappa_i \sim p\,, \tag{2.8.20}$$

and using $\kappa = \omega/c$ for the photon, Eq. (2.8.19) emerges from Eq. (2.8.20). These two equations show that in this limit, corresponding to the region of saturation, c), the photon and electron have become kinematically indistinguishable. This occurs when the linear momentum and energy transfer is such that the electron is accelerated towards c. In this condition, its momentum can no longer be related to its velocity through an equation such as $p = ?\ m_0 c$. If such a relation is tried in Eq. (2.8.19), there emerges $\hbar\omega_i = ?\ m_0 c^2$, which contradicts the initial assumption $\hbar\omega_i \gg m_0 c^2$. The electron traveling at c loses its mass, but retains the momentum $\hbar\kappa_i = p$, and so Newtonian concepts become inapplicable. This difficulty is inherent in the axioms of special relativity themselves - the concept of mass loses meaning in a particle traveling at c. The obvious conclusion is that neither the photon nor the electron can be regarded as particles without mass, and neither can travel at c, only infinitesimally near c. This conclusion is reinforced rigorously by the emergence [15] of the Evans-Vigier field $\boldsymbol{B}^{(3)}$, which implies that the photon has three degrees of space polarization. Its Wigner little group [15] is therefore *O(3)* and it cannot be massless. An electron traveling infinitesimally near c is well known [17] to be concomitant with electromagnetic plane waves which are indistinguishable at these velocities from those concomitant with the photon. A plot of electron kinetic energy verses its momentum is a constant in region c).

Region a), the quadratic region, refers in this context to very low energy photons, so $\omega_i \sim \omega_f$ in Eq. (2.8.17) and $v \ll c$ where v is the electron's speed, acquired in a collision with the incoming photon. If $\omega_i \sim \omega_f$, Eq. (2.8.17) shows that

$$m_0 c^2 \sim \left(p^2 c^2 + m_0^2 c^4\right)^{1/2}, \tag{2.8.21}$$

and so $p \ll m_0 c$, i.e., $v \ll c$ if $p = m_0 v$. Therefore the electron's kinetic energy in region a) is the non-relativistic $1/2\left(m_0 v^2\right)$, and the plot of kinetic energy against linear momentum is quadratic.

An intermediate region b) develops in which the kinetic energy is linearly proportional to its momentum. Therefore simple Compton theory parallels the qualitative features of the relativistic inverse Faraday effect. The existence of region b) can be inferred as follows. The relativistic kinetic energy is [18]

$$T = En - m_0 c^2, \tag{2.8.22}$$

where

$$En^2 = c^2 p^2 + m_0^2 c^4, \tag{2.8.23}$$

and where the relativistic momentum is $\boldsymbol{p} = \gamma_| m_0 \boldsymbol{v}$. If we assume that the rest energy, $m_0 c^2$, is small compared with cp, then

$$T \sim cp = \gamma_| m_0 vc := \left(1 - \frac{v^2}{c^2}\right)^{-1/2} m_0 vc, \tag{2.8.24}$$

so that the kinetic energy is directly proportional to the relativistic momentum, and approximately proportional to the classical momentum, $m_0 \boldsymbol{v}$, if v/c is still fairly small. The latter condition holds in region b).

8.5 Discussion

The inverse Faraday effect and related magneto-optic effects, when detected with visible frequency light, appear to be proportional to the light intensity I_0, because we are in region a), the non-relativistic limit. The nearest approach to region b) to date appears to have been the experiment of Deschamps *et al.* [7], with microwave pulses. It has been shown [19] that the conditions in this experiment correspond, for 3 GHz pulse, to

$$\omega \sim 5 \frac{e}{m_0} B^{(0)} , \tag{2.8.25}$$

and so (from Eq. (2.8.1)) we are still in region a). The experimental demonstration of the existence of region b), and of the characteristic $I_0^{1/2}$ profile of the Evans-Vigier field, requires an experiment of the type carried out by Deschamps *et al.* [7], but with a peak microwave pulse power density about two orders of magnitude greater. With contemporary technology this is entirely feasible. Related magneto-optic phenomena should also enter region b) under the right experimental conditions, i.e., high beam power density and low beam frequency. Laser sources in the visible have the opposite characteristics, i.e., high frequency compared with power density, and so produce region a), the non-relativistic limit. So magneto-optic phenomena of this kind [2] have always been seen to be proportional to I_0.

The approach to region c) appears with contemporary technology to be very difficult, because it requires enormous pulse power density in comparison to beam frequency.

Acknowledgments

It is a pleasure to acknowledge helpful discussion and correspondence with several colleagues, including: Gareth J. Evans, Stanislaw Kielich, Mikhail A. Novikov, Mark P. Silverman, Jean-Pierre Vigier, Stanisław Woźniak and Boris Yu Zel'dovich.

References

[1] S. Kielich and A. Piekara, *Acta Phys. Polon.* **18**, 439 (1959).

[2] R. Zawodny in M. W0. Evans, and S. Kielich, eds., *Modern Nonlinear Optics,* Vol. 85(1) of *Advances in Chemical Physics,* I. Prigogine and S. A. Rice, eds., (Wiley Interscience, New York, 1993).

[3] M. W. Evans, *The Photon's Magnetic Field* (World Scientific, Singapore, 1992).

[4] G. Wagnière, *Linear and Nonlinear Optical Properties of Molecules* (Verlag Helvetica Chimica Acta, Basel, 1993).

[5] P. S. Pershan, *Phys. Rev.* **130**, 919 (1963).

[6] J. P. van der Ziel, P. S. Pershan, and L. D. Malmstrom, *Phys. Rev.* **143**, 574 (1966).

[7] J. Deschamps, M. Fitaire, and M. Lagoutte, *Phys. Rev. Lett.* **25**, 1330 (1970); *Rev. Appl. Phys.* **7**, 155 (1972).

[8] S. Woźniak, M. W. Evans, and G. Wagnière, *Mol. Phys.* **75**, 81, 99 (1992).

[9] M. W. Evans, *Physica B* **182**, 227, 237 (1992).

[10] M. W. Evans, *Physica B* **183**, 103 (1993).

[11] M. W. Evans, *Mod. Phys. Lett.* **7**, 1247 (1993); *Found. Phys. Lett.* **7**, 67 (1994).

[12] A. A. Hasanein and M. W. Evans, *The Photomagneton and Quantum Field Theory* , Vol. 1 of *Quantum Chemistry* (World Scientific, Singapore, 1994).

[13] M. W. Evans in A. van der Merwe and A. Garuccio, eds., *Waves and Particles in Light and Matter* (Plenum, New York, 1994), proceedings of the de Broglie centennial.
[14] M. W. Evans, *Found. Phys.* and *Found. Phys. Lett.* in press.
[15] M. W. Evans and J.-P. Vigier, *The Enigmatic Photon, Vol. 1: The Field* $\boldsymbol{B}^{(3)}$. (Kluwer, Dordrecht, 1994); *The Enigmatic Photon, Vol. 2: Non-Abelian Electrodynamics* (Kluwer Academic, Dordrecht, 1995)
[16] P. W. Atkins, *Molecular Quantum Mechanics*, 2nd edn, Chap.1 (Oxford University Press, Oxford, 1983).
[17] J. D. Jackson, *Classical Electrodynamics* (Wiley, New York, 1962).
[18] J. B. Marion and S. T. Thornton, *Classical Dynamics of Particles and Systems*, 3rd. edn. (HBJ, Fort Worth, 1988).
[19] M. W. Evans, *Found. Phys. Lett.* in press.

Paper 9

On the Irrotational Nature of the $\boldsymbol{B}^{(3)}$ Field

Circularly polarized radiation produces phaseless magnetic effects in matter, an observation which can be explained through the fundamental $\boldsymbol{B}^{(3)}$ field of the radiation. It is shown that the irrotational nature of this field is compatible with a multipole expansion of the radiation field.

Key words. Irrotational nature of $\boldsymbol{B}^{(3)}$; multipole representation of $\boldsymbol{B}^{(3)}$

9.1 Introduction

Several magneto-optic effects are known in nature, the earliest one to be observed is the inverse Faraday effect [1—3], in which circularly polarised radiation produces a phase free magnetization similar to that produced by a static magnetic field aligned in one axis (Z). Magnetization by electromagnetic radiation has recently been interpreted [4—8] using a theorem which expresses the conjugate product of non-linear optics in terms of a phaseless magnetic field $\boldsymbol{B}^{(3)}$ which is the space component of a Pauli-Lubanski four-vector ($|\boldsymbol{B}^{(3)}|$, $\boldsymbol{B}^{(3)}$). If it is assumed that $\boldsymbol{B}^{(3)}$ is zero there is no classical field helicity [5]. Therefore $\boldsymbol{B}^{(3)}$ is a fundamental field akin to the particle helicity introduced by Wigner [10]. Vector analysis leads to the conclusion that the empirically observed conjugate product [1—3] is uniaxial if $\boldsymbol{B}^{(3)}$ is aligned with Z. In this simple case the field is therefore irrotational, its curl is zero because it is a simple axial vector in one axis: the empirical observation of the conjugate product in magneto-optics leads directly to this conclusion if we consider the conjugate product to be made up of plane waves propagating in the axis, Z, in which $\boldsymbol{B}^{(3)}$ is aligned by definition. Therefore it can be expressed in terms of the gradient of a scalar function because the curl of such a function is always identically zero. If the plane waves are replaced by components of multipole radiation, then the $\boldsymbol{B}^{(3)}$ vector in vacuo is still irrotational for all multipole components. This is shown as follows.

In Sec. 9.2 the $\boldsymbol{B}^{(3)}$ field is worked out for multipole components of the radiated electromagnetic field. It is assumed for the sake of argument that the magnetic monopole does not exist in nature, although there are data which counter-indicate this assumption [11—13]. Therefore the scalar function of which $\boldsymbol{B}^{(3)}$ is a gradient obeys Laplace's equation, whose solutions are well known in electrostatics. It is therefore straightforward to show that $\boldsymbol{B}^{(3)}$ in general can be expressed in terms of multipole components in the spherical harmonic expansion, involving, as usual, the well known Legendre polynomials.

It is emphasized that every individual component in the multipole expansion of $\boldsymbol{B}^{(3)}$ is irrotational, because every component is a particular solution of the Laplace equation for the scalar function of which $\boldsymbol{B}^{(3)}$ is a gradient both by empirical observation [1—3] and by definition. This is a direct and clear consequence of two basic premises: that $\boldsymbol{B}^{(3)}$ is phaseless (produces observable phase free magnetic effects [1—3]) and that there exist no observed magnetic monopoles in nature. If data show that magnetic monopoles exist on the contrary, the Laplace equation is replaced by a Poisson equation, with physical consequences which can be worked out with the well known solutions of Poisson's equation.

This simple line of argument has been developed in this paper in order to show that $\boldsymbol{B}^{(3)}$ in multipole radiation is irrotational for all multipole components if there are no magnetic monopoles. Recent arguments in the literature which claim that $\boldsymbol{B}^{(3)}$ is somehow not irrotational [14,15] are counter-indicated by the arguments developed here and discussed in Sec. 9.3, in which the scalar function of which $\boldsymbol{B}^{(3)}$ is a gradient is identified as a Stratton scalar potential [16—18]. The Laplacian of this scalar potential is zero if there are no magnetic monopoles in nature, and the solution of the Laplace equation allows $\boldsymbol{B}^{(3)}$ to be expanded in terms of multipoles, in precisely the same way as angular momentum. The relation between the longitudinal $\boldsymbol{B}^{(3)}$ and the transverse $\boldsymbol{B}^{(1)} = \boldsymbol{B}^{(2)*}$ in vacuo (the B cyclic theorem) is a theorem of $\hat{C}$ *negative* angular momentum components. As pointed out by Atkins [19] this allows the development of a large fraction of all quantum theory. The B cyclic theorem can therefore be used straightforwardly to quantize the electromagnetic field in vacuo.

9.2 Laplace Equation for the Gradient Function of $\boldsymbol{B}^{(3)}$

Since $\boldsymbol{B}^{(3)}$ is empirically phaseless (*i.e.*, observed in nature to be phaseless and independent of the electromagnetic frequency and wavevector) and if it is assumed that there are no magnetic monopoles (magnetic charges or sources present) then it can be expressed in terms of the gradient of a scalar function:

$$\boldsymbol{B}^{(3)} = -\nabla \Phi_B , \tag{2.9.1}$$

which is determined by the well known Laplace equation [18,20],

$$\nabla^2 \Phi_B = 0 . \tag{2.9.2}$$

If $\boldsymbol{B}^{(3)}$ depended on the electromagnetic phase, it would oscillate at high frequencies and no phaseless magneto optic effects would have been observed [1—3]. Therefore the time dependent part of $\boldsymbol{B}^{(3)}$ is zero, leading to Eq. (2.9.1). (Analogously a Coulomb field can be expressed as the gradient of a scalar potential which obeys the Laplace equation in a source free region such as the vacuum in conventional electrostatics.)

To find the general form of $\boldsymbol{B}^{(3)}$ in a multipole expansion, we therefore solve the Laplace equation for Φ_B, and evaluate the gradient of this solution, which is [19,20],

$$\Phi_B = \frac{U(r)}{r}\,\rho(\theta)\,Q(\phi)\,, \tag{2.9.3}$$

in spherical polar coordinates (r, θ, ϕ). The general solution (2.9.3) can be written as [19,20],

$$\phi_B = (Ar^l + Br^{-2})Y_{lm}(\theta, \phi)\,, \tag{2.9.4}$$

where $Y_{lm}(\theta, \phi)$ are the spherical harmonics and A and B are constants. Here m and l are integers, with l running from $-m$ to m. The solution of Laplace's equation is therefore obtained [19,20] as a product of radial and angular functions. The latter are orthonormal functions, the spherical or tesseral harmonics, which form a complete set on the surface of the unit sphere for the two indices l and m. Integer l defines the order of the multipole component, $l = 1$ is a dipole; $l = 2$ is a quadrupole; $l = 3$ is an octopole; $l = 4$ is a hexadecapole and so forth. The properties of the spherical harmonics are very well known.

The most general form of $\boldsymbol{B}^{(3)}$ from Laplace's equation is therefore,

$$\left.\begin{aligned} \boldsymbol{B}^{(3)} &= -\nabla\Phi_m\,, \\ \Phi_m &= (Ar^l + Br^{-2})Y_{lm}(\theta, \phi)\,. \end{aligned}\right\} \tag{2.9.5}$$

This is the phaseless magnetic field of multipole radiation. The solution (2.9.5) reduces to the simple [4—8],

$$\boldsymbol{B}^{(3)} = B^{(0)}\boldsymbol{e}^{(3)} = B^{(0)}\boldsymbol{k}\,,$$

when $l = 1$, $m = 0$, $r = Z$, $\theta = 0$, $A = -B^{(0)}$, $B = 0$ and $\nabla = (\partial/\partial Z)\,\boldsymbol{k}$. More generally, there exist other irrotational forms of $\boldsymbol{B}^{(3)}$,

a) The $\boldsymbol{B}^{(3)}$ for dipolar radiation, $l = 1$, $m = -1, 0, 1$.
b) The $\boldsymbol{B}^{(3)}$ for quadrupole radiation, $l = 2$, $m = -2, -1, 0, 1, 2$.
c) The $\boldsymbol{B}^{(3)}$ for octopole radiation, $l = 3$, $m = -3, -2, -1, 0, 1, 2, 3$.
d) The $\boldsymbol{B}^{(3)}$ for hexadecapole radiation, $l = 4$, $m = -4, -3, -2, -1, 0, 1, 2, 3, 4$.
e) The $\boldsymbol{B}^{(3)}$ for n pole radiation, $l = n$, $m = -n, \ldots, n$.

The $\boldsymbol{B}^{(3)}$ for *n*-pole fields are irrotational for all n and are all solutions of Maxwell's equations and generalizations such as those due to Majorana [21] and Weinberg [22], Ahluwalia *et al.* [23] and Dvoeglazov *et al.* [24]. They are all phaseless and all contribute to magneto optical effects. In every case the longitudinal and transverse components are angular momentum components expressible in the language of spherical harmonics.

9.3 Discussion

In Sec. 9.2, we have firstly used empirical evidence from magneto optics to argue that the fundamental $\boldsymbol{B}^{(3)}$ field is irrotational because for plane waves it is a simple vector defined in one axis (Z). The possible forms of $\boldsymbol{B}^{(3)}$ for n pole radiation were then worked out using the Laplace equation, *i.e.*, by expressing $\boldsymbol{B}^{(3)}$ as the negative of the gradient of a scalar function. This procedure is equivalent to using the static part of a Stratton potential for the magnetic field [25]. The complete form of the Stratton potential is [26],

$$\boldsymbol{B} = -\nabla\phi_m - \frac{1}{c}\frac{\partial \boldsymbol{A}_m}{\partial t}, \qquad (2.9.7)$$

and as shown recently by Afanasiev and Stepanofsky [27], the Stratton potential is needed for a complete description of the classical electromagnetic helicity in terms of a conservation equation and Noether's Theorem. This finding is consistent with the fact that the helicity is zero if $\boldsymbol{B}^{(3)}$ is zero, a basic inconsistency in conventional electrodynamics [20], which uses a *U(1)* gauge and asserts that $\boldsymbol{B}^{(3)}$ is zero.

It is also well known (for example problem 6.6 of Jackson's first edition [20]) that *any* vector field ($\boldsymbol{B}$) can be expressed as the sum of irrotational and divergentless components under well defined conditions. This is consistent with the fact that the transverse plane wave $\boldsymbol{B}^{(1)} = \boldsymbol{B}^{(2)*}$ is divergentless while the longitudinal $\boldsymbol{B}^{(3)}$ is irrotational. Therefore we can write:

$$\boldsymbol{B} = \boldsymbol{B}^{(1)} + \boldsymbol{B}^{(2)} + \boldsymbol{B}^{(3)}, \tag{2.9.8a}$$

$$\nabla \cdot \boldsymbol{B}^{(1)} = \nabla \cdot \boldsymbol{B}^{(2)} = \nabla \cdot \boldsymbol{B}^{(3)} = 0, \tag{2.9.8b}$$

$$\nabla \times \boldsymbol{B}^{(3)} = \boldsymbol{0}, \tag{2.9.8c}$$

$$\boldsymbol{B}^{(1)} \times \boldsymbol{B}^{(2)} = i B^{(0)} \boldsymbol{B}^{(3)*}, \quad \text{et cyclicum}, \tag{2.9.8d}$$

and the B cyclic equation (2.9.8) is a condition under which $\boldsymbol{B}^{(1)} = \boldsymbol{B}^{(2)*}$ is divergentless and $\boldsymbol{B}^{(3)}$ is both irrotational and divergentless. This is self-consistent and consistent with empirical data from magneto optics [1—3]. This result will not be found in conventional electrodynamics because the former introduces an *O(3)* gauge through Eq. (2.9.8). It is, however, consistent with Laplace's equation as shown in Sec. 9.2.

In conventional electrostatics, the Coulomb field is expressed as the negative of the gradient of a scalar function in the presence of charges (electric monopoles). The Poisson equation is then solved to show that the scalar potential is proportional to the charge density in the universe. If there is no charge density, the scalar potential is zero and there is no Coulomb field. In the Coulomb gauge therefore there is no longitudinal, irrotational electric field present in the conventional treatment of electrostatics and electrodynamics. The introduction of Maxwell's displacement current (a vacuum current) allows in electrodynamics the existence of transverse waves which are conventionally unaccompanied by the Coulomb field. Therefore the development of conventional (Maxwellian) electrodynamics is based on the existence without charges of a current, Maxwell's displacement current, made up of the time derivative of a transverse electric field which exists in the absence of sources (electric monopoles). This is self-inconsistent in several ways, as discussed recently by Chubykalo *et al. [28]* and by Lehnert *et al.* [29]. The most fundamental inconsistency is that the charge (or monopole) and the field take on a separate identity, the field, according to Maxwell, can exist without the charge, because Maxwell's displacement current can exist in the absence of sources. If so, it is equally valid, following Lehnert [29] to assume that the divergence of the electric field is non-zero in the absence of sources, or to introduce the vacuum convection current, following Chubykalo *et al.* [28]. Both procedures lead directly to $\boldsymbol{B}^{(3)}$ in the vacuum. Furthermore, the B cyclic theorem (2.9.8d) is a relation bewteen field components in the absence of magnetic monopoles, *i.e. in vacuo*, or *in the vacuum*.

A point of major importance, and a turning point in the development of electrodynamics, is that magneto-optical data have now been identified as giving direct and unequivocal empirical support for the existence of $\boldsymbol{B}^{(3)}$ and an *O(3)* gauge. This is also logical support for Lehnert *et al.* [29] and for Chubykalo *et al.* [28], who have developed recently a self-consistent form of electrodynamics. Ultimately, it seems logical to develop electrodynamics and unified field theory on the basis that the primordial field exists in the vacuum, following Maxwell, and to assume that charge is a manifestation of the field as originally supposed by Faraday [28]. It also seems possible [28] to develop a fully covariant theory in electrodynamics which allows velocities greater than c and which allows the interrelation of field theory with action at a distance theory [28]. Proceeding on this basis, the B cyclic theorem becomes the archetypical theorem of the primordial vacuum field. It is simply a relation between components of spin angular momentum multiplied by a $\hat{C}$ negative coefficient. Thus, the electromagnetic field is a physical entity which has angular and linear momentum, as observed empirically in nature.

Finally, if we assume that magnetic monopoles exist in the universe, the fundamental magnetic field $\boldsymbol{B}^{(3)}$ can be expressed as the negative of the gradient of a scalar function which is the solution of a Poisson equation,

$$\Phi_B(\boldsymbol{r}) = \Phi_B(0) + \frac{1}{4\pi}\int \frac{\nabla' \cdot \boldsymbol{B}(\boldsymbol{r}', t)}{|\boldsymbol{r} - \boldsymbol{r}'|}\, d^3r' . \tag{2.9.9}$$

Here $\Phi_B(0)$ is a constant of integration, as discussed by Jackson [20] on his page 8 of the first edition, and where the divergence of the complete magnetic field $\boldsymbol{B} = \boldsymbol{B}^{(1)} + \boldsymbol{B}^{(2)} + \boldsymbol{B}^{(3)}$ is non-zero because magnetic multipoles are assumed to be present in the universe. The $\boldsymbol{B}^{(3)}$ field from this solution is axial, conservative and irrotational, in precise analogy to the central Coulomb field in the presence of electric monopoles in the universe.

It is concluded that the fundamental $\boldsymbol{B}^{(3)}$ field responsible for magneto optical effects is irrotational both in the absence and in the presence of magnetic monopoles. As described on Jackson's page nine of the first edition [20], the line integral of Stokes' Theorem $\oint \boldsymbol{B}^{(3)} \cdot d\boldsymbol{l}$ is zero over any closed path. This is a counter argument to Comay's recent assertion [14] that $\boldsymbol{B}^{(3)}$ is not irrotational. Clearly, if this were true, $\boldsymbol{B}^{(3)}$ would not be proportional to the empirically observable conjugate product appearing in the B cyclic theorem (2.9.8d), and $\boldsymbol{B}^{(3)}$ would not be a fundamental C negative angular momentum of the electromagnetic field in vacuo: the primordial and fundamental electromagnetic spin angular momentum. As such, $\boldsymbol{B}^{(3)}$ remains irrotational for n pole radiation and for plane waves in the presence and absence of electric and magnetic monopoles.

Acknowledgments

The Alpha Foundation is warmly thanked for an affiliation and many colleagues for interesting Internet discussions.

References

[1] J. P. van der Ziel, P. S. Pershan, and L. D. Malmstrom, *Phys. Rev. Lett.* **15**, 190 (1965); *Phys. Rev.* **143**, 574 (1966).
[2] J. Deschamps, M. Fitaire, and M. Lagoutte, *Phys. Rev. Lett.* **25** 1330 (1970); *Rev. Appl. Phys.* **7**, 155 (1972).
[3] R. Zawodny in M. W. Evans and S. Kielich, eds., *Modern Nonlinear Optics* Vol. 85(1) of *Advances in Chemical Physics*, I Prigogine and S.Rice, eds. (Wiley Interscience, New York, 1997, paperback, 3rd printing), a review of magneto optics with ca. 150 references.
[4] M. W. Evans, *Physica B* **182**, 227, 237 (1992).
[5] M. W. Evans and J.-P. Vigier, *The Enigmatic Photon, Volume 1: The Field* $\boldsymbol{B}^{(3)}$ (Kluwer Academic, Dordrecht, 1994).
[6] M. W. Evans and J.-P. Vigier, *The Enigmatic Photon, Volume 2: Non-Abelian Electrodynamics* (Kluwer Academic, Dordrecht, 1995).
[7] M. W. Evans, J.-P. Vigier, S. Roy and S. Jeffers, *The Enigmatic Photon, Volume 3: Theory and Practice of the* $\boldsymbol{B}^{(3)}$ *Field* (Kluwer Academic, Dordrecht, 1996).
[8] M. W. Evans, J.-P. Vigier, and S. Roy, eds., *The Enigmatic Photon, Volume 4: New Directions* (Kluwer Academic, Dordrecht, 1998), with review articles from leading specialists.
[9] M. W. Evans and A. A. Hasanein, *The Photomagneton in Quantum Field Theory* (World Scientific, Singapore, 1994).
[10] E. P. Wigner, *Ann. Math.* **40**, 149 (1939).
[11] M. Israelit, *Magnetic Monopoles and Massive Photons in a Weyl Type Electrodynamics* LANL Preprint 9611060 (1996), *Found. Phys.*, in press.
[12] B Barish, G. Lin, and C. Lane, *Phys. Rev. D* **36**, 2641 (1987).
[13] N. Rosen, *Found. Phys.* **12**, 213 (1982).

[14] E. Comay, *Chem. Phys. Lett.* **261**, 601 (1996).
[15] M. W. Evans and S. Jeffers, *Found. Phys. Lett.* **9**, 587 (1996), reply to Ref. 15.
[16] A. F. Ranada, *Eur. J. Phys.* **13**, 70 (1992).
[17] A. F. Ranada, *J. Phys. A* **25**, 1621 (1992).
[18] H. Bacry, *Helv. Phys. Acta* **67**, 632 (1994).
[19] P. W. Atkins, "*Molecular Quantum Mechanics*, 2nd edn., (Oxford Univ. Press, Oxford, 1983).
[20] J. D. Jackson, *Classical Electrodynamics* (Wiley, New York, 1962).
[21] E. Majorana, *Nuovo* Cim. **14**, 171 (1937); papers in the Domus Galileiana, Pisa; *E. Gianetto, Lett. Nuovo Cim.* **44**, 140 (1985), theory developed in ref. (8) by E. Recami and M. W. Evans.
[22] S. Weinberg, *Phys. Rev.* **133B**, 1318 (1964); 134B, 882 (1964).
[23] D. V. Ahluwalia and D. J. Ernst, *Mod. Phys. Lett* **7A**, 1967 (1992).
[24] V. V. Dvoeglazov, *Int. J. Theor. Phys.* **35**, 115 (1996); V. V. Dvoeglazov, Yu. N. Tyukhtyaev and S. V. Khudyakov, *Russ. J. Phys* **37**, 898 (1994), *Rev. Mex. Fis. Suppl.* **40**, 352 (1994); "Majorana Like Models in the Physics of Neutral Particles.", Int. Conf. Theory Electron, Cuautitlan, Mexico, 1995; in Ref.(8, "The Weinberg Formalism and New Looks at Electromagnetic Theory.", a review with ca. 100 references.
[25] H. K. Moffat, *Nature* **347**, 367 (1990).
[26] H. Pfister and W. Gekelman, *Am. J. Phys.* **59**, 497 (1994).
[27] G. N. Afanasiev and Yu. P. Stepanovsky, "The Helicity of the Free Electromagnetic Field and its Physical Meaning." Preprint E2-95-413, J.I.N.R., Dubna, 1995; Nuovo Cim., in press. Also M. W. Evans, *Physica A* **214**, 605 (1995); V. V. Dvoeglazov, "On the Claim that the Antisymmetric Field Tensor is Longitudinal after Quantization." in Ref. 8.
[28] A. E. Chubykalo and R. Smirnov-Rueda, *Phys. Rev. E* **53**, 5373 (1996); ibid., Ref. 8, "Action at a Distance and Self-Consistency of Classical Electrodynamics.", a review; ibid., "The Convective Displacement Current." (World Scientific, planned); A. E. Chubykalo, M. W. Evans, and R. Smirnov-Rueda, *Found. Phys. Lett.* **10**, 93 (1997).

[29] B. Lehnert, *Phys. Scripta* **53**, 204 (1996); *Optik* **99**, 113 (1995); in Ref. 8, "Electromagnetic Space-Charge Waves in Vacuo.", a review; B. Lehnert and S. Roy, "Electromagnetic Theory with Space Charges in Vacuo and Non-Zero Rest Mass of Photon." (World Scientific, planned)

Paper 10

The Interaction of the Evans-Vigier Field with Atoms

The theoretical study is initiated of the interaction of the Evans-Vigier field, $\boldsymbol{B}^{(3)}$, with atomic matter, represented by atoms such as H in which there is net electronic spin angular momentum only, and by atoms in which there is both spin and orbital electronic angular momentum. In H it is inferred from the Dirac equation that the net spin of the ground state electron should interact directly with $\boldsymbol{B}^{(3)}$, so that the Zeeman splitting due to such an interaction should be proportional to the *square root* of the power density of the beam, whatever its frequency. In atoms or molecules in which there is also net orbital angular momentum, the effect of $\boldsymbol{B}^{(3)}$ at first and second order is treated approximately using the classical Hamilton-Jacobi equation.

Key words: Evans-Vigier field, atomic interaction with.

10.1 Introduction

It has been shown recently [1,2] that the interaction of the classical electromagnetic field with one electron is governed entirely by the Evans-Vigier field, $\boldsymbol{B}^{(3)}$. Thus far, this has been shown in two ways, using the classical but relativistic Hamilton-Jacobi equation for the orbital electronic angular momentum, and the Dirac equation for the spin electronic angular momentum, which has no classical meaning [3]. Other methods of demonstration can be used, for example using the Dirac equation in its Hamilton-Jacobi form. If the newly inferred $\boldsymbol{B}^{(3)}$ were zero, there would be no observable interaction between the electron and the field, contrary to experience. It is well known experimentally [4] that intense microwave pulses magnetize an electron plasma, a process which the classical Hamilton-Jacobi equation ascribes *entirely* to $\boldsymbol{B}^{(3)}$, acting at first and second orders in $B^{(0)}$, the scalar magnitude of the magnetic flux density of the circularly polarized microwave pulse. For sufficiently low frequencies and intense pulses, the magnetization is dominant at first order in B^{0}, and is therefore proportional to $I_0^{1/2}$, where I_0 is the power density of the pulse in W m^{-2}.

If for simplicity we consider the interaction of one electron with the electromagnetic field in relativistic quantum field theory [3], the Dirac equation shows [1] that the permanent magnetic dipole moment ($\boldsymbol{m}$) set up by the intrinsic electronic spin forms an interaction Hamiltonian $H_s = -\boldsymbol{m} \cdot \boldsymbol{B}^{(3)}$. The magnetization due to $\boldsymbol{m}$ is $N\boldsymbol{m}$ if there are N electrons in a plasma. It is a permanent magnetization whose average value is zero, not one induced by the electromagnetic field. In atoms such as H however, this type of interaction Hamiltonian leads to an optical Zeeman effect for all beam frequencies, and Sec. 10.2 is an account of the fundamentals of this effect, which is detectible with electron spin resonance [5]. Section 10.3 extends the discussion to atoms (and molecules) with net orbital as well as spin electronic angular momentum. Finally, a discussion is given in the Born-Oppenheimer approximation of fundamental magnetic dipole transitions involving $\boldsymbol{B}^{(3)}$.

10.2 Optical Zeeman Effect in H Due to $\boldsymbol{B}^{(3)}$

In atomic H, the single electron is bound to the nucleus in an orbital, and its net orbital angular momentum is quenched to zero [6]. It therefore has only a net spin angular momentum ($\boldsymbol{S}$) from the Dirac equation. In a conventional uniform magnetic field, the Zeeman effect occurs due to the electronic spin angular momentum, and is essentially an observable splitting of the atomic absorption spectrum of H [7]. In a free electron, on the other hand, there are no atomic absorption lines, and the Zeeman effect in an electron plasma is not observable in this way. In order to understand the interaction of electromagnetic radiation with atomic or free electrons, the Dirac equation is necessary, because without it, there is no spin quantum number S. Therefore we begin our discussion with a summary of the Dirac equation for one electron (e) in the electromagnetic field represented by the four-potential A_μ, and explain the emergence [1] of the Evans-Vigier field from the first principles of relativistic quantum field theory. Thereafter the discussion is extended to the Dirac equation of the electron of the H atom in the electromagnetic field, with emphasis on $\boldsymbol{B}^{(3)}$. This should be regarded as only the first step towards the rigorous understanding of the interaction of $\boldsymbol{B}^{(3)}$ with atomic matter. A fuller and more detailed understanding will rely on numerical methods, because the Dirac equation becomes analytically intractable.

The Dirac equation for the interaction of the free electron with the electromagnetic field is [1]

$$\gamma_\mu \left(p_\mu + eA_\mu\right)\psi\left(\boldsymbol{p}\right) = -m_0 c\psi\left(\boldsymbol{p}\right), \tag{2.10.1}$$

where γ_μ is the Dirac matrix [3] and ψ the Dirac four-spinor. In Minkowski notation,

$$\gamma_\mu = \left(\gamma, i\gamma^{(0)}\right), \qquad A_\mu = \left(A, \frac{i\phi}{c}\right),$$
$$p_\mu = \left(p, \frac{iEn}{c}\right), \tag{2.10.2}$$

and m_0 is the electron's mass. Using standard methods of solution [1] Eq. (2.10.1) reduces to the energy eigenequation

$$\hat{W}u = Hu, \tag{2.10.3}$$

where u is a Dirac four-spinor in the standard representation [1—3], and where the Hamiltonian eigenvalue H is

$$H = \frac{1}{2m_0}\left(\sigma\cdot\left(\boldsymbol{p} + e\boldsymbol{A}\right)\right)^2 - e\phi\,. \tag{2.10.4}$$

Here σ is a Pauli spinor [1—3]. It can be shown [1] that the part of H that describes the interaction of the electron's intrinsic spin with the electromagnetic field is

$$H_s = \frac{i\sigma}{2m_0}\cdot\left(\boldsymbol{p} + e\boldsymbol{A}\right)\times\left(\boldsymbol{p} + e\boldsymbol{A}\right) = \frac{e\hbar}{2m_0}\sigma\cdot\boldsymbol{B}^{(3)}$$
$$= -\boldsymbol{m}\cdot\boldsymbol{B}^{(3)}, \tag{2.10.5}$$

where $\boldsymbol{B}^{(3)}$ is the Evans-Vigier field [8—12] of vacuum electromagnetism. Here $\boldsymbol{S} = (\hbar\sigma)/2$ is the electronic spin angular momentum, and e/m_0 the charge to mass ratio of the electron. Therefore $\boldsymbol{B}^{(3)}$ is to the vacuum electromagnetic field as $\boldsymbol{S}$ is to the electron, an irremovable component. If $\boldsymbol{B}^{(3)}$ were zero, then $\boldsymbol{S}$ could not interact with the electromagnetic field, in contradiction with the structure of the Dirac equation itself.

To extend this analysis to the H atom requires an extra term V in the Hamiltonian describing the fact that the electron is bound in an orbital to the nucleus. In the Born-Oppenheimer approximation [13] the Hamiltonian is split into a part dealing with the isolated atom (i.e., the way in which the electron is bound to the nucleus) and an interaction term. The latter can be expressed in terms of atomic property tensors, such as the magnetic dipole moment of the one free electron,

$$\boldsymbol{m} = -\frac{e}{2m_0}\boldsymbol{S} . \tag{2.10.6}$$

In this approximation, the interaction of the electron in the H atom with the applied electromagnetic field is described in the same way as that for the free electron of Eq. (2.10.4), i.e., through the dot product of $\boldsymbol{m}$ and the $\boldsymbol{B}^{(3)}$ of circularly polarized electromagnetic radiation in vacuo. In the H atom, however, there are observable electronic spectra in the visible and ultra-violet regions of the electromagnetic range of frequencies [14], spectra which appear as discrete absorption or emission lines. These do not occur in the electron plasma, because they are due essentially to atomic structure [7]. These spectral features are now known with great precision, and can be used to measure the effect of the $\boldsymbol{B}^{(3)}$ field through its Zeeman effect [7].

The Zeeman effect of $\boldsymbol{B}^{(3)}$ in atomic H occurs because the Dirac equation is based on spinors, which imply that the eigenvalues of the H atom's electronic spin angular momentum are $\hbar/2$ and $-\hbar/2$. In $\boldsymbol{B}^{(3)}$ these are no longer degenerate, and transitions between them are possible. These were first measured in atomic H in an ordinary magnetic field by Beringer and Heald [15] about forty years ago, and should be measurable for $\boldsymbol{B}^{(3)}$ with contemporary technology. If so, the $\boldsymbol{B}^{(3)}$ Zeeman effect should depend on $I_0^{1/2}$ because $\boldsymbol{B}^{(3)}$ depends on $I_0^{1/2}$. In the same way as the ordinary Zeeman effect, due to an ordinary magnetic field, depends on its flux density, so does the Zeeman effect due to the Evans-Vigier field, which has all the known properties [1,2] of a magnetic field. We refer to this predicted phenomenon as the optical Zeeman effect. In atomic H it should be measurable with the use of pulses of high intensity circularly polarized pump radiation and contemporary synchronized detection. For example the ESR signal of H should be shifted by $\boldsymbol{B}^{(3)}$, and the shift should be proportional to $I_0^{1/2}$. Similar optical Zeeman effects should be observable in other atoms with net ground state electron spin, such as the alkali metal vapors. (The original Zeeman effect was observed in sodium vapor.) In order to observe them it is necessary to use circularly polarized pulses of very high intensity, because $\boldsymbol{B}^{(3)}$ is zero [1] in linearly or incoherently polarized radiation, and because

$$B^{(0)} = \left(\frac{I_0}{\epsilon_0 c^3} \right)^{\frac{1}{2}}, \tag{2.10.7}$$

where ϵ_0 is the vacuum permittivity and c the speed of light in vacuo. Therefore for a power density of 1.0 W cm^{-2}, (10^4 W m^{-2}), the magnitude of $B^{(0)}$ is only about 10^{-5} T, ten times smaller than the Earth's mean magnetic field, producing a very small Zeeman shift.

Since $\boldsymbol{B}^{(3)}$ emerges from the Dirac equation itself, it is non-zero, and is an observable. If definitive experimental evidence to the contrary is obtained, then the Dirac equation will have failed at a basic level.

10.3 Atoms and Molecules with Orbital Electronic Angular Momentum

The equilibrium of one electron in the electromagnetic field can be considered classically with the relativistic Hamilton-Jacobi equation [1,2]. Consider the condition,

$$\omega = \frac{e}{m_0} B^{(0)}, \tag{2.10.8}$$

where e/m_0 is the charge to mass ratio of the electron and where $B^{(0)}$ is the scalar magnitude of the magnetic flux density of the beam. Under condition (2.10.8), ω is both the angular frequency of the beam and the orbital angular frequency of the electron in equilibrium with the beam. For the electromagnetic beam, ω is κc where κ is the magnitude of the wave-vector, and where c is the speed of light. For the electron, the de Broglie matter-wave equation [1] gives

$$\omega = \frac{1}{\hbar}\left(m_0^2 c^4 + \hbar^2\kappa^2 c^2\right)^{\frac{1}{2}} . \tag{2.10.9}$$

Under condition (2.10.8), using Eq. (2.10.9), we obtain

$$m_0^2 c^4 + \hbar^2\kappa^2 c^2 = \hbar^2 \frac{e^2}{m_0^2} B^{(0)2} , \tag{2.10.10}$$

which is a *cyclotron condition* for equilibrium of the electron in the field. In the limit,

$$\kappa \gg \frac{m_0 c}{\hbar} , \tag{2.10.11}$$

(where κ now refers to the matter wave of the electron) we obtain the result,

$$\omega \sim \kappa c \sim \frac{eB^{(0)}}{m_0} , \tag{2.10.12}$$

which, using the relation between $A^{(0)}$ and $B^{(0)}$ of the wave,

$$A^{(0)} = \frac{B^{(0)}}{\kappa} , \tag{2.10.13}$$

is the charge quantization condition [1]. The electron and photon become indistinguishable in the limit (2.10.11) in which the cyclotron frequency of the electron is the angular frequency of the beam.

This simple illustration shows that the beam can be thought of as driving the electron in an orbit, which can be described in terms of classical, relativistic mechanics. This is quite different from the interaction process of the intrinsic spin angular momentum with the field, a process which has no classical interpretation. In an atom in which the electron has both orbital and spin angular momentum, the Zeeman splitting pattern due to $\boldsymbol{B}^{(3)}$ is therefore affected by both types of angular momentum appearing in the appropriate Hamiltonian operator of the Dirac equation. We expect phenomena to order $I_0^{1/2}$ and I_0 in the optical Zeeman spectrum in such atoms, and in general, these can be understood only by solving the Dirac equation numerically.

A qualitative understanding of the problem can be attained, however, by considering an atom with one electron, an electron which has orbital as well as spin angular momentum, and by splitting off the interaction Hamiltonian of the electron with the field from the Hamiltonian describing the way in which the electron is bound to the nucleus. This type of approximation is the basis of semi-classical radiation theory [13], in which atomic property tensors are treated quantum mechanically, and the field classically. In the non-relativistic approximation several predicted phenomena of the Evans-Vigier field have been described [8—12]. In the fully relativistic treatment, however, major new features emerge — at microwave frequencies the interaction Hamiltonian (and therefore the $\boldsymbol{B}^{(3)}$ induced Zeeman splitting) becomes proportional to $B^{(0)}$, and therefore to I_0. At visible frequencies the process is dominated by the term in I_0, because the classical Hamilton-Jacobi equation of the free electron in the field shows these properties. This result explains why shifts caused by visible lasers of atomic frequencies appear experimentally to be dominated [16] by an I_0 dependence. It is clear that the new theoretical understanding provided by $\boldsymbol{B}^{(3)}$ implies the need for a fundamental re-appraisal of the interpretation of spectra such as these.

10.4 Discussion

The existence of $\boldsymbol{B}^{(3)}$ in vacuo means that there must be a novel magnetic dipole interaction, which is not considered in the usual Born-Oppenheimer approximation. In the usual semi-classical approach to radiation theory [13], the latter leads to interaction Hamiltonians which include magnetic dipole terms such as

$$H_{int} = -\boldsymbol{m} \cdot \boldsymbol{B}(t) . \tag{2.10.14}$$

In this expression, however, $\boldsymbol{B}(t)$ is time dependent and originates in the plane wave $\boldsymbol{B}^{(1)} = \boldsymbol{B}^{(2)*}$, not in the $\boldsymbol{B}^{(3)}$ field itself. The novel interaction term $H_s = -\boldsymbol{m} \cdot \boldsymbol{B}^{(3)}$,, however, is *fundamental* in atomic and molecular spectroscopy whenever a circularly polarized field is used, i.e., whenever the Evans-Vigier field is non-zero in vacuo. Some effects of H_s have been discussed in Secs. 10.2 and 10.3, and should be observable with contemporary technology. Magnetization by circularly polarized electromagnetic radiation [1] also depends on an interaction term of this type, and this can be understood classically for the free electron interacting with the field through the relativistic Hamilton-Jacobi equation. The Dirac equation in its Hamilton-Jacobi form [17], or an equivalent quantum equation, must be used to extend this understanding to atoms, so that a fully consistent theory emerges in relativistic quantum mechanics. It is already clear, however, that if $\boldsymbol{B}^{(3)}$ were zero, the Hamilton-Jacobi equation itself would give a meaningless conclusion.

Acknowledgments

It is a pleasure to acknowledge many interesting ideas put forward in discussions with several colleagues, among these were: Yildirim Aktas, Gareth J. Evans, Ahmed A. Hasanein, the late Stanislaw Kielich, Mikhail Novikov, Mark P. Silverman, Jean-Pierre Vigier, and B. Yu. Zel'dovich.

References

[1] M. W. Evans and J.-P. Vigier, *The Enigmatic Photon, Volume 1: The Field* $\boldsymbol{B}^{(3)}$ (Kluwer, Dordrecht, 1994); *The Enigmatic Photon, Volume 2: Non-Abelian Electrodynamics.* (Kluwer Academic, Dordrecht, 1995).

[2] M. W. Evans, *Found. Phys. Lett.* in press, 1994.

[3] L. H. Ryder, *Quantum Field Theory* 2nd edn. (Cambridge University Press, Cambridge, 1987).

[4] J. Deschamps, M. Fitaire, and M. Lagoutte, *Phys. Rev. Lett.* **25**, 1330 (1970); *Rev. Appl. Phys.* **7**, 155 (1972).

[5] S. Woźniak, M. W. Evans and G. Wagnière, *Mol. Phys.* **75**, 81, 99 (1992).

[6] P. W. Atkins, *Molecular Quantum Mechanics* 2nd edn. (Oxford University Press, Oxford, 1983).

[7] M. W. Evans, *Spec. Sci. Tech.* **16**, 43 (1993).

[8] A. A. Hasanein and M. W. Evans, *The Photomagneton and Quantum Field Theory* (World Scientific, Singapore, 1994), Vol. 1 of *Quantum Chemistry.*

[9] M. W. Evans, and S. Kielich, eds., *Modern Nonlinear Optics*, Vol. 85(2) of *Advances in Chemical Physics*, I. Prigogine and S. A. Rice, eds., (Wiley Interscience, New York, 1993).

[10] M. W. Evans, *Physica B* **182**, 227, 237 (1992); **183**, 103 (1993); **190**, 310 (1993).

[11] M. W. Evans, *The Photon's Magnetic Field.* (World Scientific, Singapore, 1992).

[12] M. W. Evans, *Found. Phys. Lett.* **7**, 67 (1994); *Mod. Phys. Lett.* **7**, 1247 (1993).

[13] G. H. Wagnière, *Linear and Nonlinear Optical Properties of Molecules.* (Verlag Helvetica Chimica Acta, Basel, 1993), Appendix 1.

[14] B. W. Shore and D. H. Menzel, *Principles of Atomic Spectra* (Wiley, New York, 1968).

[15] R. Beringer and M. A. Heald, *Phys. Rev.* **95**, 1474 (1955).

[16] W. Happer, *Rev. Mod. Phys.* **44**, 169 (1972).

Paper 11

The Derivation of the Majorana Form of Maxwell's Equations from the B Cyclic Theorem

It is demonstrated that the *B* Cyclic theorem (equivalence principle) of the new electrodynamics gives Majorana's form of Maxwell's equations in the vacuum. This demonstration provides a link between the new and received views of vacuum electrodynamics, showing that the equations of motion can be derived from the equivalence principle, assuming only the correspondence principle of quantum mechanics. Therefore the *B* Cyclic theorem is quantized to give the Maxwell equations in Majorana's form.

11.1 Introduction

In the past few years it has become clear that a major advance in electrodynamics has occurred. The electromagnetic field is now thought to have longitudinal components in the vacuum [1—7], one of which, conveniently referred to as the $\boldsymbol{B}^{(3)}$ field, being phase free and observable empirically, for example in magneto-optics. This longitudinally directed magnetic field forms part of the structure of the B Cyclic theorem [1—3], which inter-relates transverse and longitudinal components in the vacuum. (There is also an equivalent theorem in the presence of sources and matter.) The purpose of this Letter is to show that the Maxwell equations in the form given by Majorana [8—10] can be derived from the B Cyclic theorem, showing that Maxwell's equations themselves must give longitudinal solutions which are inter-related to the usual transverse electromagnetic waves through a novel principle of equivalence between space-time and the electromagnetic field.

11.2 Derivation of the Maxwell Equations from the *B* Cyclic Theorem

The equivalence principle of the new electrodynamics, the B Cyclic theorem, inter-relates the transverse and longitudinal components of the vacuum electromagnetic field as follows,

$$\boldsymbol{B}^{(1)} \times \boldsymbol{B}^{(2)} = iB^{(0)}\boldsymbol{B}^{(3)*}, \qquad \text{et cyclicum,} \tag{2.11.1}$$

where $\boldsymbol{B}^{(1)} = \boldsymbol{B}^{(2)*}$ is the transverse component (e.g. a plane wave) and where

$$\boldsymbol{B}^{(3)} = B^{(0)}\boldsymbol{e}^{(3)}, \tag{2.11.2}$$

is the longitudinal, phase free, component in the basis ((1), (2), (3)), a complex basis of the rotation sub-group *O(3)* of the Poincaré group. For the present purposes it proves convenient to reduce Eq. (2.11.1) to cyclics in the vector potential $\boldsymbol{A}$, defined in *S.I.* units by

$$\boldsymbol{B} = \nabla \times (i\boldsymbol{A}), \tag{2.11.3a}$$

$$\boldsymbol{E} = c\nabla \times \boldsymbol{A}, \tag{2.11.3b}$$

where $\boldsymbol{B}$ is magnetic flux density, $\boldsymbol{E}$ is electric field strength, and c the speed of light in vacuo. Equation (2.11.3) uses the Hertz-Stratton representation of $\boldsymbol{E}$ as the curl of an axial (*i.e.*, rotational) $\boldsymbol{A}$, for example the plane wave,

$$\boldsymbol{A}^{(1)} = \frac{A^{(0)}}{\sqrt{2}}(i\boldsymbol{i} + \boldsymbol{j})\,e^{i\phi}, \qquad \phi = \omega t - \kappa Z. \tag{2.11.4}$$

Here $\boldsymbol{e}^{(1)} = (\boldsymbol{i} - i\boldsymbol{j})/\sqrt{2}$; $A^{(0)}$ is a scalar amplitude, and ϕ is the electromagnetic phase, where ω is the angular frequency at instant t, and κ the wavevector at Z. Using Eq. (2.11.3) and (2.11.4) gives the standard result [1—7],

$$\boldsymbol{E}^{(1)} = \frac{E^{(0)}}{\sqrt{2}}(\boldsymbol{i} - i\boldsymbol{j})e^{i\phi},$$

$$\boldsymbol{B}^{(1)} = \frac{B^{(0)}}{\sqrt{2}}(i\boldsymbol{i} + \boldsymbol{j})e^{i\phi}, \tag{2.11.5}$$

from which it is inferred that $\boldsymbol{A}$ is axial and $i\boldsymbol{A}$ is polar. Self consistently, therefore, the complex polar vector $\boldsymbol{E}$ is the curl of the complex axial vector $\boldsymbol{A}$; and the complex axial vector $\boldsymbol{B}$ is the curl of the complex polar vector $i\boldsymbol{A}$. Since $\boldsymbol{A}$ is axial, it is described by the $\boldsymbol{A}$ cyclics:

$$\boldsymbol{A}^{(1)} \times \boldsymbol{A}^{(2)} = iA^{(0)}\boldsymbol{A}^{(3)*}, \tag{2.11.6a}$$

$$\boldsymbol{A}^{(2)} \times \boldsymbol{A}^{(3)} = iA^{(0)}\boldsymbol{A}^{(1)*}, \tag{2.11.6b}$$

$$\boldsymbol{A}^{(3)} \times \boldsymbol{A}^{(1)} = iA^{(0)}\boldsymbol{A}^{(2)*}, \tag{2.11.6c}$$

where

$$\begin{aligned} B^{(0)} &= \kappa A^{(0)}, & E^{(0)} &= -ic\kappa A^{(0)}, \\ \boldsymbol{B}^{(1)} &= \kappa \boldsymbol{A}^{(1)}, & \boldsymbol{E}^{(1)} &= -ic\kappa \boldsymbol{A}^{(1)}, \\ \boldsymbol{B}^{(2)} &= \kappa \boldsymbol{A}^{(2)}, & \boldsymbol{E}^{(2)} &= -ic\kappa \boldsymbol{A}^{(2)}, \\ \boldsymbol{B}^{(3)} &= \kappa \boldsymbol{A}^{(3)}, & \boldsymbol{E}^{(3)} &= -ic\kappa \boldsymbol{A}^{(3)}. \end{aligned} \tag{2.11.7}$$

Now multiply both sides of Eq. (2.11.6a) by i and transform to Cartesian coordinates to give

$$\left(iA^{(0)}\right)A_Z^* + i\left(\left(iA_X\right)A_Y^* - \left(iA_Y\right)A_X^*\right) = 0, \tag{2.11.8}$$

and it is seen that this equation has the Cartesian structure,

$$p^{(0)}\psi_Z + i\left(p_X\psi_Y - p_Y\psi_X\right) = 0, \tag{2.11.9}$$

where $\boldsymbol{p}$ is a polar vector, $\boldsymbol{\psi}$ an axial vector, and $p^{(0)}$ the scalar magnitude of $\boldsymbol{p}$. Equation (2.11.9) is one of the Maxwell equations as derived by Majorana [8—10]. In the vacuum, the axial vector potential is defined by

$$\boldsymbol{A}^{(1)} = \boldsymbol{A}^{(2)*} = \frac{c}{\kappa}\left(c\boldsymbol{B}^{(1)} + i\boldsymbol{E}^{(1)}\right), \tag{2.11.10}$$

and the polar vector potential $i\boldsymbol{A}$ is identified through the correspondence principle with the del operator,

$$\frac{1}{e}\boldsymbol{p} = i\boldsymbol{A} \rightarrow -i\frac{\hbar}{e}\hat{\nabla}. \tag{2.11.11}$$

The $iA^{(0)}$ function is identified with a time differential operator through the same correspondence principle, i.e.,

$$\frac{1}{e}p^{(0)} = iA^{(0)} \rightarrow i\frac{\hbar}{ec}\frac{\hat{\partial}}{\partial t}. \qquad (2.11.12)$$

Similarly the other two equations of the A cyclics reduce to the other two Majorana-Maxwell equations as follows,

$$\boldsymbol{A}^{(2)} \times \boldsymbol{A}^{(3)} = iA^{(0)}\boldsymbol{A}^{(1)*} \rightarrow p^{(0)}\psi_X + i\left(p_Y\psi_Z - p_Z\psi_Y\right) = 0, \qquad (2.11.13)$$

and

$$\boldsymbol{A}^{(3)} \times \boldsymbol{A}^{(1)} = iA^{(0)}\boldsymbol{A}^{(2)*} \rightarrow p^{(0)}\psi_Y + i\left(p_Z\psi_X - p_X\psi_Z\right) = 0, \qquad (2.11.14)$$

11.3 Discussion

The cyclic equations of the new electrodynamics reduce to the Majorana form of the Maxwell equations using the correspondence principle in the form (2.11.11) and (2.11.12), which is also a form of the minimal prescription for the free field [1—7], i.e., the proportionality between linear momentum and the vector potential, the latter having the dimensions of linear momentum multiplied by charge. The novel cyclic field equations represent an equivalence principle between rotation generators of *O(3)* and the electromagnetic field. Using the correspondence principle, the equivalence principle reduces to Maxwell's equations in the form given by Majorana, in which complex field combinations take the role of wavefunctions. The Maxwell equations have therefore been *derived* from a more fundamental cyclical structure, and have therefore been derived in a form which has *O(3)* symmetry. This is the Majorana form of Maxwell's equations in vacuo.

Acknowledgments

Many colleagues worldwide are acknowledged for Internet discussions, and in particular, Professor Erasmo Recami is thanked for sending relevant preprints, and a copy of *Il Caso Majorana*, his best-selling scientific biography of Ettore Majorana. York University, Toronto is thanked for a visiting professorship.

References

[1] M. W. Evans and J.-P. Vigier, *The Enigmatic Photon, Vol. 1: The Field* $\boldsymbol{B}^{(3)}$ Kluwer Academic, Dordrecht, 1994).

[2] M. W. Evans and J.-P. Vigier, *The Enigmatic Photon, Vol. 2: Non-Abelian Electrodynamics* (Kluwer Academic, Dordrecht, 1995).

[3] M. W. Evans, J.-P. Vigier, S. Roy, and S. Jeffers, *The Enigmatic Photon, Vol. 3: Theory and Practice of the* $\boldsymbol{B}^{(3)}$ *Field* (Kluwer Academic, Dordrecht, 1996).

[4] M. W. Evans, J.-P. Vigier, and S. Roy, eds., *The Enigmatic Photon, Vol. 4: New Developments* (Kluwer, Dordrecht, in prep), a collection of contributed papers.

[5] M. W. Evans and S. Kielich, eds., *Modern Nonlinear Optics,* Vol. 85(2) of *Advances in Chemical Physics,* I. Prigogine and S. A. Rice, eds. (Wiley Interscience, New York, 1993).

[6] A. A. Hasanein and M. W. Evans, *The Photomagneton in Quantum Field Theory* (World Scientific, Singapore, 1994).

[7] M. W. Evans, *The Photon's Magnetic Field* (World Scientific, Singapore, 1992).

[8] R. Mignani, E. Recami, and M. Baldo, *Lett. Il Nuovo Cim.* **11**, 568 (1974).

[9] E. Giannetto, *Lett. Il Nuovo Cim.* **44**, 140, 145 (1985).

[10] Erasmo Recami, *Il Caso Majorana* (Bestsellers Saggi, Milan, 1991, paperback).

Paper 12

The Evans-Vigier Field $\boldsymbol{B}^{(3)}$ Interpreted as a De Broglie Pilot Field

A straightforward consideration of the antisymmetric part of the tensor of free space light intensity leads to the result $\boldsymbol{B}^{(3)} / B^{(0)} = \boldsymbol{J}^{(3)} / \hbar$, where $\boldsymbol{B}^{(3)}$ is the Evans-Vigier field [1—10], a phase free magnetic flux density of amplitude $B^{(0)}$ carried in free space by the electromagnetic wave component, and where $\boldsymbol{J}^{(3)} = \hbar \boldsymbol{e}^{(3)}$, with $\boldsymbol{e}^{(3)}$ being a unit axial vector in the propagation axis. The field $\boldsymbol{B}^{(3)}$ therefore pilots the photon angular momentum, $\boldsymbol{J}^{(3)}$. The consequences are discussed of the wave-particle duality inherent in this result, using diffraction patterns due to $\boldsymbol{B}^{(3)}$ in a double slit interferometer.

Key words: $\boldsymbol{B}^{(3)}$ Field, de Broglie pilot field.

12.1 Introduction

It has been inferred recently [1—10] that the conventional view of free space electromagnetism is incomplete, because the classical wave interpretation produces a novel phase free magnetic flux density in the vacuum, the Evans-Vigier field $\boldsymbol{B}^{(3)}$. The latter exists in free space because there exists the *electromagnetic torque* *density* $iB^{(0)}\boldsymbol{B}^{(3)*}/\mu_0 = \boldsymbol{B}^{(1)} \times \boldsymbol{B}^{(2)}/\mu_0$. Here μ_0 is the vacuum permeability in *S.I.* units [11] and $B^{(0)}$ the scalar amplitude of the magnetic flux density produced in free space by Maxwell's equations. In this notation, the usual energy density in free space is the dot product of complex plane waves [12],

$$U = \frac{1}{\mu_0}\boldsymbol{B}^{(1)} \cdot \boldsymbol{B}^{(2)}, \tag{2.12.1}$$

in $J\ m^{-3}$, and we work in a complex representation [13] of three dimensional space, a representation defined by the cyclically symmetric unit vector algebra,

$$\boldsymbol{e}^{(1)} \times \boldsymbol{e}^{(2)} = i\boldsymbol{e}^{(3)*}. \tag{2.12.2}$$

Since scalar light intensity, I_0 ($W\ m^{-2}$), is, in *S.I.* units [14],

$$I_0 = cU, \tag{2.12.3}$$

the *imaginary* axial vector quantity,

$$\boldsymbol{I}_A = \frac{c}{\mu_0}\boldsymbol{B}^{(1)} \times \boldsymbol{B}^{(2)} = i\frac{c}{\mu_0}B^{(0)}\boldsymbol{B}^{(3)*}, \tag{2.12.4}$$

is the antisymmetric part of the free space light intensity tensor [5]. So the imaginary $\boldsymbol{I}_A$ is directly proportional to the *real* and physical Evans-Vigier field $\boldsymbol{B}^{(3)}$ in vacuo [6].

In this Letter, it is shown that the equation,

$$\frac{\boldsymbol{B}^{(3)}}{B^{(0)}} = \frac{\boldsymbol{J}^{(3)}}{\hbar} = \boldsymbol{e}^{(3)}, \tag{2.12.5}$$

is a straightforward consequence of the quantization of the electromagnetic field. In Eq. (2.12.5), $\hbar$ is the Dirac constant, and the real and physical

$$\boldsymbol{J}^{(3)} = \hbar \boldsymbol{e}^{(3)}, \tag{2.12.6}$$

is an angular momentum with magnitude $\hbar$ of the field as particle. In Sec. 12.2, the result (2.12.5) is derived straightforwardly from fundamentals. In Sec. 12.3, diffraction patterns are discussed qualitatively, patterns due to $\boldsymbol{B}^{(3)}$ of a circularly polarized wave passing through a double slit interferometer [15]. Finally, a discussion is pursued of the field-particle duality inherent in Eq. (2.12.5), in that $\boldsymbol{B}^{(3)}$ and $\boldsymbol{J}^{(3)}$ are directly proportional. Since $\boldsymbol{B}^{(3)}$ is produced from a cross product of vector plane wave functions $\boldsymbol{B}^{(1)}$ and $\boldsymbol{B}^{(2)}$, it satisfies the criteria originally proposed [16] by de Broglie for pilot waves. Since $\boldsymbol{B}^{(3)}$ is phase free and is a magnetic flux density, it is referred to henceforth as the *pilot field* for $\boldsymbol{J}^{(3)} = \hbar \boldsymbol{e}^{(3)}$, and the usual term *wave-particle* is replaced by *field-particle*.

12.2 Derivation of Equation (2.12.5) from Fundamentals

Torque has the same units as energy and is the time derivative of angular momentum. Therefore, there exist in vacuum electromagnetism *torque densities* (i.e., torques per unit volume),

$$\boldsymbol{T}_V^{(3)*} := -\frac{1}{\mu_0}\boldsymbol{B}^{(1)} \times \boldsymbol{B}^{(2)} = -i\frac{B^{(0)}}{\mu_0}\boldsymbol{B}^{(3)*}, \text{ et cyclicum,} \tag{2.12.7}$$

in which

$$\boldsymbol{m}^{(1)} = \frac{\boldsymbol{B}^{(1)}}{\mu_0}, \tag{2.12.8}$$

and so on are oscillating magnetic dipole moments of the radiation itself. Thus,

$$\boldsymbol{T}_V^{(3)*} = -\boldsymbol{m}^{(1)} \times \boldsymbol{B}^{(2)}, \tag{2.12.9}$$

in formal analogy with the definition of magnetically generated torque in electrostatics and electrodynamics [12]. However, the *real* (i. e. physical) part of $\boldsymbol{T}_V^{(3)*}$ is identically zero because the real, physical, angular momentum density, $\boldsymbol{J}_V^{(3)}$, of the beam in vacuo is constant. Thus,

$$Re\left(\boldsymbol{T}_V^{(3)*}\right) = \frac{\partial}{\partial t}\boldsymbol{J}_V^{(3)} = \boldsymbol{0}. \tag{2.12.10}$$

Now use in Eq. (2.12.10) one of the standard axioms of quantum mechanics, one based on the de Broglie relation, the axiom

$$\frac{\partial}{\partial t} = -i\frac{En}{\hbar}, \tag{2.12.11}$$

where En is energy. Since $\boldsymbol{J}_V^{(3)}$ is real, Eqs. (2.12.7), (2.12.10) and (2.12.1) give an imaginary

$$\boldsymbol{T}_V^{(3)*} = -i\frac{En}{\hbar}\boldsymbol{J}_V^{(3)*} = -i\frac{En}{\hbar V}\boldsymbol{J}^{(3)*}, \tag{2.12.12}$$

where V is the volume used to define $\boldsymbol{J}_V^{(3)*}$, and where the real $\boldsymbol{J}^{(3)*}$ now has the units of angular momentum itself rather than angular momentum density. In vacuum electromagnetic radiation, the energy density En/V is given by Eq. (2.12.1),

$$U = \frac{En}{V} = \frac{B^{(0)2}}{\mu_0}, \tag{2.12.13}$$

and so

$$\boldsymbol{T}_V^{(3)*} = -i\frac{B^{(0)}}{\mu_0}\boldsymbol{B}^{(3)*} = -i\frac{B^{(0)2}}{\mu_0\hbar}\boldsymbol{J}^{(3)*}, \qquad (2.12.14)$$

from which

$$\boldsymbol{B}^{(3)} = B^{(0)}\frac{\boldsymbol{J}^{(3)}}{\hbar}, \qquad (2.12.15)$$

which is Eq. (2.12.5), with $\boldsymbol{J}^{(3)} = \hbar\boldsymbol{e}^{(3)}$. The result (2.12.15), or (2.12.5), was first derived in Ref. 1, using another method, and re-derived independently in Ref. 6.

Equation (2.12.5) can be derived in another way using a straightforward adaptation of the standard expression for $\hbar\omega$ in quantum field theory [5], the Planck-Einstein light quantum hypothesis,

$$\hbar\omega = \int U dV. \qquad (2.12.16)$$

Instead of the usual $U = \boldsymbol{B}^{(1)} \cdot \boldsymbol{B}^{(2)}/\mu_0$,, we use

$$U = \frac{|\boldsymbol{B}^{(1)} \times \boldsymbol{B}^{(2)}|}{\mu_0} = \frac{B^{(0)2}}{\mu_0}, \qquad (2.12.17)$$

and obtain

$$\hbar = \frac{1}{\mu_0\omega}\int |\boldsymbol{B}^{(1)} \times \boldsymbol{B}^{(2)}| \, dV. \qquad (2.12.18)$$

In the basis (2), Eq. (2.12.18) becomes

$$iJ^{(3)*} := i\hbar e^{(3)*} = \frac{iB^{(0)}}{\mu_0\omega}\int B^{(3)*}\,dV, \tag{2.12.19}$$

and rearranging,

$$B^{(3)} = \frac{\mu_0\omega}{B^{(0)}V}J^{(3)} = B^{(0)}\frac{J^{(3)}}{\hbar}, \tag{2.12.20}$$

which is again Eq. (2.12.5).

It has been shown that Eq. (2.12.5) is the direct, self-consistent result of two fundamental axioms of quantum mechanics, Eqs. (2.12.11) and (2.12.16).

12.3 Diffraction Patterns Due to $B^{(3)}$

In order to understand quantitatively the implications of Eq. (2.12.5) in field-particle duality, it is necessary to consider a Young experiment for $B^{(3)}$ carried out with a *circularly polarized* incident beam, which is diffracted through the double aperture of the interferometer to form a diffraction pattern. In classical electromagnetism, this requires an exact solution of the Maxwell equations as described recently by Jeffers *et al.* [17] for linearly polarized incident radiation. For linearly polarized radiation, however, $B^{(3)}$ nets to zero, because it changes sign from right to left circular polarization [6]. It would therefore be interesting to repeat the work of Jeffers *et al.* [17] for circularly polarized incident radiation and to map the $B^{(3)}$ diffraction patterns quantitatively and accurately. In the absence of such data we draw a qualitative sketch of the patterns to be expected using the relation between the magnitude of $B^{(3)}$ and beam intensity (I_0, in W m^{-2}),

$$\boldsymbol{B}^{(3)} = B^{(0)}\boldsymbol{e}^{(3)} = \left(\frac{I_0}{\epsilon_0 c^3}\right)^{1/2} \boldsymbol{e}^{(3)}, \tag{2.12.21}$$

where ϵ_0 is the *S.I.* vacuum permittivity. Equation (2.12.21) expresses $\boldsymbol{B}^{(3)}$ in terms of $I_0^{1/2}$, and magnetization [11,18] due to $\boldsymbol{B}^{(3)}$ therefore has an $I_0^{1/2}$ profile which is observable in principle using microwave pulses to magnetize an electron plasma [18]. Fig. (9) of Jeffers *et al.* [17] shows lines of constant I_0 forming a diffraction pattern indistinguishable from an *interferogram* one which shows considerable structure [17] within a few wavelengths of the slits. This structure is unobtainable [17] in the usual scalar theory of diffraction [12], and the energy flow is calculated along paths normally interpreted as an interference pattern. However, as pointed out by Jeffers *et al.* [17], there is no such thing present as a classical interference, i.e., no radiation actually crosses the axis of symmetry. In this view, no radiation passing through the top aperture arrives at a point below the axis of symmetry and vice-versa.

Qualitatively, we expect similar patterns for the diffracted $\boldsymbol{B}^{(3)}$ to be determined by Eq. (2.12.21) through the *square root* of the intensity. The significance of Eq. (2.12.21), and of the expected $\boldsymbol{B}^{(3)}$ diffraction pattern, is discussed as follows.

12.4 Discussion

Although this exact, classical analysis [17] of diffraction appears to be in need of extension to incident circular polarization, in which $\boldsymbol{B}^{(3)}$ is non-zero, we discuss here the inference that $\boldsymbol{B}^{(3)}$ is the pilot field of $\hbar$, the photon's angular momentum. If so, $\boldsymbol{B}^{(3)}$ and $\hbar$ are simultaneously measurable in the de Broglie-Vigier-Bohm interpretation [19] of the quantum theory. Lines of constant $\boldsymbol{B}^{(3)}$ in a diffraction pattern would be lines of constant $\hbar\boldsymbol{e}^{(3)}$ in the basis (2). These ideas do not occur in conventional electrodynamics [12] in which $\boldsymbol{B}^{(3)}$ is undeveloped. The existence of $\boldsymbol{B}^{(3)}$ in vacuo [6], however, has by now been demonstrated in many ways, two of which are given for the first time in this Letter. The magnetizing effect of $\boldsymbol{B}^{(3)}$ can be demonstrated [6] using the classical Hamilton-Jacobi equation of one electron (e) in the classical electromagnetic field represented by the four-potential (A_μ), a demonstration which shows that the trajectory of the electron in the beam is governed *entirely* by $\boldsymbol{B}^{(3)}$ and by no other vacuum field. In *retrospect* it has become clear that this is due to the fact that $\boldsymbol{B}^{(1)} \times \boldsymbol{B}^{(2)} / \mu_0$ is a torque density of radiation in the vacuum, and to the fact that $\boldsymbol{B}^{(3)}$ is directly proportional to the radiation's angular momentum density (Eq. (2.12.5)). Prior to this however, $\boldsymbol{B}^{(1)} \times \boldsymbol{B}^{(2)}$ was an almost unknown quantity labelled by some nonlinear opticians as the *conjugate product*, although it is only one out of several possible conjugate products of the vacuum electromagnetic field [5—11]. The obscurity of this language precluded its clear interpretation, but the magnetic conjugate product is the radiation torque density multiplied by the vacuum permeability, a torque density which is simply $iB^{(0)}\boldsymbol{B}^{(3)*}/\mu_0$. The torque per unit volume of radiation is therefore directly proportional to the real and physical Evans-Vigier field $\boldsymbol{B}^{(3)*} \left(= \boldsymbol{B}^{(3)} \right)$ of electromagnetism in the vacuum.

The use of the classical Hamilton-Jacobi equation of e in A_μ to demonstrate the existence of $\boldsymbol{B}^{(3)}$ from the principle of least action [6] is significant in at least two ways. Firstly, in a historical context, Cushing [20] has pointed out that de Broglie originally saw the classical Hamilton-Jacobi equation as providing "...an embryonic theory of the union of waves and particles, all in a manner consistent with a realist conception of matter". Equation (2.12.5) now shows that if $\hbar$ is the angular momentum of a particle, the photon, then $\hbar$ must be directly linked with $\boldsymbol{B}^{(3)}$, and in the realist view, be simultaneously observable with it. Rewriting Eq. (2.12.5),

$$\boldsymbol{J}^{(3)} = \hbar \left(\frac{\boldsymbol{B}^{(3)}}{B^{(0)}} \right), \tag{2.12.22}$$

we obtain an expression which is directly analogous with the Planck-Einstein and de Broglie relations, $En = \hbar\omega$ and $\boldsymbol{p} = \hbar\boldsymbol{\kappa}$ respectively.

Secondly, as shown by Bohm [21], the Schrödinger equation can be interpreted in the realist manner [22] by developing it into a quantized Hamilton-Jacobi equation, provided that the quantum potential is introduced, and provided that non-locality and concepts such as superluminal action at a distance are accepted as valid hypothesis. (The causal, realist point of view of Selleri *et al.* [23] does not accept action at a distance.) These questions are addressed in the interesting volume [7] recording the de Broglie centennial.

In the Copenhagen agreement [7] on the other hand, the quantum equivalent of $\boldsymbol{B}^{(3)}$ is interpreted as an angular momentum operator, $\hat{B}^{(3)}$, the photomagneton [1,4,6]. The latter is directly proportional to $\hat{J}^{(3)}$ in the vacuum, and $\hat{J}^{(3)}$ is an angular momentum operator of quantum mechanics in the Copenhagen view, and obeys the commutator relations of such operators. Is it possible to use the Evans-Vigier field to distinguish between the Copenhagen and realist interpretations of quantum mechanics? In order to begin to scratch the surface of this question, we can adapt Bohm's original discussion [2] as far as possible, in the context of diffraction patterns for $\boldsymbol{B}^{(3)}$. The line of argument is that if $\boldsymbol{B}^{(3)}$ is the de Broglie pilot wave of $\hbar$, it is simultaneously measurable with $\hbar$ in the realist view.

In the classical theory of electrodynamics [12], $\boldsymbol{B}^{(3)}$ is expected to be modified by diffraction as described accurately for the first time by Jeffers *et al.* [17]. This is a purely classical phenomenon which can be inferred by solving Maxwell's equations with the appropriate boundary conditions. The diffraction patterns after passing through the double apertures are those of $\boldsymbol{B}^{(3)}$ itself, so must be those of the angular momentum of the radiation. The latter can be represented after quantization by $\hbar\boldsymbol{e}^{(3)}$, whose magnitude is $\hbar$. The particle (photon) concomitant with the diffracted wave therefore has angular momentum magnitude $\hbar$. If so, however, where is the particle after diffraction [7]? If the incoming wave-particle is equivalent to one photon, what happens to the photon on diffraction? This question is answered entirely differently in the realist and Copenhagen views of quantum mechanics [7]. In the simpler case of light passing through a beam divider, the duality of $\hbar$ and $\boldsymbol{B}^{(3)}$ is described as follows.

In the realist interpretation [7] the photon carries particulate information, and at random goes to one of two detectors, A or B, after the light has been split by the beam divider. However, the spin field $\boldsymbol{B}^{(3)}$ goes to both detectors simultaneously, detectors which measure split beam intensity. Since single photon (and neutron) generators are now available [7], these assumptions are experimentally explorable, and Aspect *et al.* [24] appear to have shown that if a photon goes one way, there is no photon present in the other arm, there is 100% anticorrelation. Therefore the photon angular momentum, $\hbar$, if detected by A, cannot be detected simultaneously by B. (We can imagine A and B to be ultra-sensitive absorption spectrometers that can detect the absorption of $\hbar$ through atomic or molecular selection rules [14] on angular momentum.) Therefore, if $\boldsymbol{B}^{(3)}$ is simultaneously measurable with $\hbar$, then the presence of $\boldsymbol{B}^{(3)}$ at B should be simultaneously measurable with $\hbar$ arriving at A. Since $\boldsymbol{B}^{(3)}$ is proportional to the square root of intensity, then this experiment should be feasible and would show that $\boldsymbol{B}^{(3)}$ is the de Broglie pilot field [7] for $\hbar$. In this view, the wave function is the classical, Maxwellian, wave itself, and can exist simultaneously at A and B when there is one photon at A. In this interpretation, however, it is necessary to assert that the relation between electromagnetic energy density and intensity, the classical Eq. (2.12.3), holds if the empty wave containing $\boldsymbol{B}^{(3)}$ is to be observable as intensity, i.e., power per unit area. This has to be true in the *absence* of $\hbar$.

In the Copenhagen interpretation, as described by Croca [25], the light incident on the beam splitter is divided into two wave packets, and when one of these hits a detector, A, for example, the photon has chosen that particular path. The wave function is a wave of probability, and the detector A is a measuring device which has the effect of bringing the photon into observational reality. Thus, *causality is lost* and the wave of probability at B vanishes and is lost to the physical, or measurable, world. Reality in the Copenhagen view is something that *follows* measurement, and this is counter intuitive. (However intuition is by its very nature, subjective.) Therefore the photon if detected at A has been brought into measurable existence at A, and is not measurable at B simultaneously. Therefore, the wave function of the photon, when detected at A, brings it into existence at A, but before that, the photon *exists* only as a probability. In standard quantum optics, if a photon exists at A, $\boldsymbol{B}^{(3)}$ exists at A in terms of photon creation and annihilation operators [3—6]. If there is no photon at B, (i.e., if no photon is *created* at B by the measuring device) there is no $\boldsymbol{B}^{(3)}$ at B, and so nothing at all should be detectible at B, while at A we detect $\boldsymbol{B}^{(3)}$ through the square root of intensity and $\hbar$ through our ultra-sensitive spectrometric device.

This experiment appears to be a clear way of distinguishing between these two interpretations. Evidently, if $\boldsymbol{B}^{(3)}$ is an *empty field* at B, as the realists assert [7], then it must carry *classical* intensity, even though it is supposed not to carry *quantized* energy. This point of view can be sustained logically only if the intensity of an empty wave is not a function of $\hbar\omega$, the quantum of light energy known as the photon. The reason is that there is no photon at B, while there is still intensity at B in the realist point of view.

Finally, we discuss briefly the idea of $\boldsymbol{B}^{(3)}$ as a pilot field. As discussed by Bohm [21] and Vigier [26], there is an entity, ψ, guiding the particle in the wave-particle duality of de Broglie, an entity which is written in terms of the real mechanical action S as

$$\psi = R \exp\left(i \frac{S}{\hbar} \right), \tag{2.12.23}$$

so that R^2 is the *probability* that a particle of mass m have a velocity $\boldsymbol{v} = \nabla S/m$. In his paper of 1952 [21], Bohm showed that this is a plausible idea if taken to its logical conclusion, and met the objections of Pauli to de Broglie's initial proposal, published in 1930 [27]. A slight extension of the pilot wave idea is to write

$$\psi^{(1)} = \boldsymbol{R}^{(1)} e^{(iS)/\hbar} = \psi^{(2)*}, \tag{2.12.24}$$

where S is the electromagnetic action [6,21],

$$S = \hbar(\omega t - \kappa \cdot \boldsymbol{r}), \tag{2.12.25}$$

and where $\boldsymbol{R}^{(1)}$ and $\boldsymbol{R}^{(2)}$ are in the complex circular basis (2),

$$\boldsymbol{R}^{(1)} = \boldsymbol{e}^{(1)}, \qquad \boldsymbol{R}^{(2)} = \boldsymbol{e}^{(2)}. \tag{2.12.26}$$

In this picture,

$$\psi^{(0)2} = |\psi^{(1)} \times \psi^{(2)}|, \tag{2.12.27}$$

is the probability of finding a particle with an angular momentum given by

$$|\boldsymbol{J}^{(3)}| = \frac{\partial S}{\partial \phi}, \tag{2.12.28}$$

where [21], ϕ is the azimuthal angle. Therefore $\boldsymbol{B}^{(3)}$, which is directly proportional [1,6] to $\boldsymbol{J}^{(3)}$, is a pilot field of the particulate angular momentum, $\hbar$, of the photon.

Acknowledgments

During the course of preparation, valuable discussions were pursued with several colleagues, who are acknowledged here for their help. Among these are: Robert M. Compton, Gareth J. Evans, Mikhail Novikov, Sisir Roy, Mark P. Silverman, Jean-Pierre Vigier, and Boris Yu Zel'dovich.

References

[1] M. W. Evans, *Physica B* **182**, 227, 237 (1992).
[2] M. W. Evans, *Physica B* **183**, 103 (1993); **190**, 310 (1993).
[3] M. W. Evans, *The Photon's Magnetic Field* (World Scientific, Singapore, 1992).

[4] A. A. Hasanein and M. W. Evans, *The Photomagneton and Quantum Field Theory* (World Scientific, Singapore, 1994).
[5] M. W. Evans and S. Kielich, eds., *Modern Nonlinear Optics*, Vols. 85(1), 85(2), 85(3) of *Advances in Chemical Physics*, I. Prigogine and S. A. Rice, eds., (Wiley Interscience, New York, 1993).
[6] M. W. Evans and J.-P. Vigier, *The Enigmatic Photon, Volume 1: The Field* $\boldsymbol{B}^{(3)}$ (Kluwer Academic Publishers, Dordrecht, 1994); ibid., *The Enigmatic Photon, Volume 2: Non-Abelian Electrodynamics* (Kluwer Academic Publishers, Dordrecht, 1995).
[7] A. van der Merwe and A. Garuccio, eds., *Waves and Particles in Light and Matter* (Plenum, New York, 1994), the de Broglie Centennial Volume.
[8] M. W. Evans, *Mod. Phys. Lett.* **7**, 1247 (1993).
[9] M. W. Evans, *Found. Phys. Lett.* **7**, 67 (1994).
[10] M. W. Evans, *Found. Phys. Lett.*, in press, 1994 / 1995.
[11] M. W. Evans, *Found. Phys.*, in press, 1994 / 1995.
[12] J. D. Jackson, *Classical Electrodynamics* (Wiley, New York, 1962).
[13] reviewed, for example, by R. Zawodny in Ref. 5, Vol. 85(1).
[14] P. W. Atkins, *Molecular Quantum Mechanics*, 2nd edn. (Oxford University Press, Oxford, 1983).
[15] reviewed by M. Bozic in Ref. 7, pp. 171 ff.
[16] reviewed by A. Garuccio in Ref. 7, pp. 37 ff.
[17] S. Jeffers, R. D. Prosser, G. Hunter and J. Sloan, Ref. 7, pp. 309 ff.
[18] J. Deschamps, M. Fitaire, and M. Lagoutte, *Phys. Rev. Lett.* **25**, 1330 (1970); *Rev. Appl. Phys.* **7**, 155 (1972).
[19] sometimes known as *The Paris School*, there are several articles in Ref. 7 devoted to the realist interpretation, e.g. E. J. Squires, pp. 125 ff.
[20] J. T. Cushing, Ref. 7, pp. 223 ff.
[21] D. Bohm, *Phys. Rev.* **85**, 166 (1952); ibid., pp. 180 ff. Bohm's refutation of Pauli's criticism of de Broglie is given in Appendix B of the second of these well known papers.
[22] A. O. Barut, in Ref. 7, pp. 9 ff.
[23] F. Selleri, Ref. 7, pp. 439 ff.
[24] P. Grangier, G. Roger and A. Aspect, *Europhys. Lett.*, **1**, 173 (1986).

[25] J. R. Croca, in Ref. 7, pp. 209 ff.
[26] J.-P. Vigier, *Present Experimental Status of the Einstein-de Broglie Theory of Light*, in *Proceedings, 4th International Symposium on Foundations of Quantum Mechanics* M. Tsukada *et al.*, eds. (Japanese Journal of Applied Physics, Tokoyo, 1993).
[27] L. de Broglie, *An Introduction to the Study of Wave Mechanics* (Dutton, New York, 1930).

Paper 13

The Charge Quantization Condition: Link Between the O(3) Gauge Group and the Dirac Equation

The charge quantization condition (*CQC*) equates the quantized vacuum photon momentum to the classical product $eA^{(0)}$, where e is the charge on the electron and where $A^{(0)}$ is the scalar magnitude of the potential four-vector of electromagnetic radiation. It is shown that the *CQC* emerges consistently from the expression for the Evans-Vigier field $\boldsymbol{B}^{(3)}$ in the *O(3)* gauge group of vacuum electromagnetism and the Dirac equation for the spinning trajectory of an electron in the field.

Key words: charge quantization condition, $\boldsymbol{B}^{(3)}$ field

13.1 Introduction

The magnetic components of the ordinary plane waves of vacuum electromagnetism are now known [1-10] to act as the source of the magnetizing field $\boldsymbol{B}^{(3)}$, the Evans-Vigier field [6]. The real and physical $\boldsymbol{B}^{(3)}$ field propagates through the vacuum with the plane waves, and is an axial vector directed in the propagation axis. It is an experimental observable, and can be isolated [6,9] from the concomitant plane waves through its magnetization of material matter, in the simplest instance one electron. The magnetization, $\boldsymbol{M}^{(3)}$, is, at microwave frequencies [11], proportional to $I_0^{1/2}$, where I_0 is the power density of the beam in W m^{-2}. Therefore $\boldsymbol{B}^{(3)}$ is a physical magnetic flux density, and is now understood in several different ways [9]. There is no reasonable doubt that it adds a third dimension to the understanding of vacuum electromagnetism.

An immediate consequence is that the gauge group of vacuum electromagnetism can no longer be considered to be the conventional *O(2)* [12], the group of rotations in a plane. The natural generalization to *O(3)*, the group of rotations in three dimensional space, is considered in Sec. 13.2, where it is shown that the field $\boldsymbol{B}^{(3)}$ emerges from *O(3)* gauge geometry as being proportional to the vector product of the plane wave vector potential $\boldsymbol{A}^{(1)}$ with its own complex conjugate $\boldsymbol{A}^{(2)}$. This result leads to the charge quantization condition (*CQC*), which equates the quantized vacuum photon momentum $\hbar\kappa$ to the classical $eA^{(0)}$. Here e is both the charge on the electron and the scaling constant of *O(3)* gauge geometry [12], and $A^{(0)}$ is the scalar magnitude of $\boldsymbol{A}^{(1)}$. In Sec. 13.3, the Dirac equation of one electron in the electromagnetic field is used to produce an expression for $\boldsymbol{B}^{(3)}$ which becomes identical with that derived in Sec. 13.2 by using the *CQC*. The latter therefore makes the *O(3)* gauge group theory of vacuum electromagnetism consistent with the Dirac equation of one electron in the electromagnetic field. Both theories consistently produce $\boldsymbol{B}^{(3)}$ in the vacuum, and a discussion is given of some of the wider implications of the discovery of the Evans-Vigier field.

13.2 The O(3) Symmetry of Vacuum Electromagnetism

The need for an *O(3)* gauge group of vacuum electromagnetism is revealed by the defining Lie algebra of the $\boldsymbol{B}^{(3)}$ field [6],

$$\begin{aligned} \boldsymbol{B}^{(1)} \times \boldsymbol{B}^{(2)} &= iB^{(0)}\boldsymbol{B}^{(3)*}, \\ \boldsymbol{B}^{(2)} \times \boldsymbol{B}^{(3)} &= iB^{(0)}\boldsymbol{B}^{(1)*}, \\ \boldsymbol{B}^{(3)} \times \boldsymbol{B}^{(1)} &= iB^{(0)}\boldsymbol{B}^{(2)*}, \end{aligned} \tag{2.13.1}$$

where $\boldsymbol{B}^{(1)} = \boldsymbol{B}^{(2)*}$ are the magnetic components of the ordinary plane waves. This algebra is *non*-Abelian, compact and semi-simple, and has *O(3)* symmetry [12], not *O(2)*. Therefore the *O(3)* group must be used to describe vacuum electromagnetism in the general theory of gauge geometries [12], a theory which parallels general relativity in its conceptual development. The *O(3)* theory of vacuum electromagnetism is non-Abelian in nature, and therefore the field can act as its own source [6]. Thus, the conjugate product $\boldsymbol{B}^{(1)} \times \boldsymbol{B}^{(2)}$ acts as the source of $\boldsymbol{B}^{(3)}$, a new physical field which propagates through the vacuum with the plane waves, and which is observed through its $I_0^{1/2}$ profile [9]. This inference is reinforced conclusively [9] because the source of $\boldsymbol{B}^{(3)}$ can be described in terms both of a Biot-Savart-Ampère law and as the curl of a vector potential [9]. Therefore $\boldsymbol{B}^{(3)}$ has all the known properties of a magnetic flux density, and acts experimentally as such [6]. In retrospect its existence has already been detected experimentally in second order magneto-optic effects, because the well known conjugate product [13] is $iB^{(0)}\boldsymbol{B}^{(3)*}$, an experimental observable. Here $B^{(0)}$ is the scalar magnitude of $\boldsymbol{B}^{(3)}$. These phenomena include: 1) the inverse Faraday effect [14]; 2) the optical Faraday effect [15]; 3) light shifts in atomic spectra induced by a circularly polarized laser at visible frequencies [16]; 4) magnetization at second order in $B^{(0)}$ of an electron plasma [17] with high intensity microwave pulses.

In field-particle physics, the general theory of gauge geometries is well developed [12], and there is a need only to adapt it for the emergence of $\boldsymbol{B}^{(3)}$ in vacuum electrodynamics. The theory is developed [12] in terms of isospin indices in an abstract isospin space whose symmetry, however, is *O(3)*. By applying this theory to the physical space (1), (2) and (3) of circular indices in which Eqs. (2.13.1) are written, the *O(3)* electromagnetic field tensors emerge [6],

$$\begin{aligned} \left(\boldsymbol{G}^{(1)*}\right)_{\mu\nu} &= \left(\boldsymbol{F}^{(1)*}\right)_{\mu\nu} - i\frac{e}{\hbar}\left(\boldsymbol{A}^{(2)} \times \boldsymbol{A}^{(3)}\right)_{\mu\nu}, \\ \left(\boldsymbol{G}^{(2)*}\right)_{\mu\nu} &= \left(\boldsymbol{F}^{(2)*}\right)_{\mu\nu} - i\frac{e}{\hbar}\left(\boldsymbol{A}^{(3)} \times \boldsymbol{A}^{(1)}\right)_{\mu\nu}, \\ \left(\boldsymbol{G}^{(3)*}\right)_{\mu\nu} &= \left(\boldsymbol{F}^{(3)*}\right)_{\mu\nu} - -i\frac{e}{\hbar}\left(\boldsymbol{A}^{(1)} \times \boldsymbol{A}^{(2)}\right)_{\mu\nu}. \end{aligned} \tag{2.13.2}$$

These generalize the usual $F_{\mu\nu}$ tensor [12] to include cross products of vector potentials. The cross product $\boldsymbol{A}^{(1)} \times \boldsymbol{A}^{(2)}$, for example, is not considered in the usual definition of $F_{\mu\nu}$ in the *O(2) (=U(1))* gauge group for electromagnetism, but is nevertheless *non-zero*, even in that gauge group, because [6]

$$\boldsymbol{B}^{(3)*} = -i\frac{\kappa}{A^{(0)}}\boldsymbol{A}^{(1)} \times \boldsymbol{A}^{(2)}. \tag{2.13.3}$$

This reveals a fundamental inconsistency in the *O(2)* gauge symmetry. In the *O(3)* gauge group, on the other hand, we obtain, self-consistently from Eq. (2.13.2),

$$\boldsymbol{B}^{(3)*} = -i\frac{e}{\hbar}\boldsymbol{A}^{(1)} \times \boldsymbol{A}^{(2)}. \tag{2.13.4}$$

Comparison of Eqs. (2.13.3) and (2.13.4) gives the charge quantization condition

$$eA^{(0)} = \hbar\kappa, \tag{2.13.5}$$

whose consistency within field theory is shown in the next section.

13.3 The Dirac Equation of One Electron in the Field

It is well known that the electron has intrinsic spin ($\boldsymbol{S}$), which has no classical meaning. This is a result of the Dirac equation recounted on numberless occasions. It has been shown recently, however, that the interaction Hamiltonian formed between $\boldsymbol{S}$ and the electromagnetic field is [6]

$$H_{spin} = \boldsymbol{S} \cdot \boldsymbol{B}^{(3)} = \frac{e\hbar\sigma}{2m_0} \cdot \boldsymbol{B}^{(3)}, \tag{2.13.6}$$

where $e/(2m_0)$ is the gyromagnetic ratio and σ is a Pauli spinor and is governed exclusively by $\boldsymbol{B}^{(3)}$, and by no other field component. Therefore $\boldsymbol{B}^{(3)}$ is to vacuum electromagnetism as $\boldsymbol{S}$ is to the electron, an intrinsic component which is not only non-zero, but irremovable. In other words, without $\boldsymbol{B}^{(3)}$, the ineluctably and characteristically quantum mechanical part of the Dirac equation of the electron in the field would be entirely and incorrectly missing. The Dirac Hamiltonian eigenvalue would become identical with the classical Hamiltonian of the electron in the field.

Thus, if $\boldsymbol{S}$ be accepted, so must $\boldsymbol{B}^{(3)}$.

The specific expression for $\boldsymbol{B}^{(3)}$ from the Dirac equation can be written as a vector cross product [6,12],

$$\boldsymbol{B}^{(3)*} = -\frac{i}{\hbar}\boldsymbol{p}^{(1)} \times \boldsymbol{A}^{(2)}, \tag{2.13.7}$$

or as a commutator of a transverse momentum *operator* $\hat{p}^{(1)}$ with the field vector potential $\boldsymbol{A}^{(2)}$. The magnetic flux density appearing in the spin part of the Hamiltonian, H_{spin}, is independent of time, and is therefore $\boldsymbol{B}^{(3)}$, because the plane waves $\boldsymbol{B}^{(1)} = \boldsymbol{B}^{(2)*}$ are time dependent and vanish on averaging at order one in $B^{(0)}$. The electron's intrinsic spin must interact directly with $\boldsymbol{B}^{(3)}$ of the field. This is a fundamental result from the first principles of relativistic quantum field theory, and cannot be discounted as a modeling procedure. Thus $\boldsymbol{B}^{(3)}$ is the fundamental magnetizing field of electromagnetic radiation at all frequencies. It is an experimental observable, whose presence in vacuo can be detected through the $I_0^{1/2}$ dependence mentioned in the introduction.

Equation (2.13.7), defining $\boldsymbol{B}^{(3)}$ from the Dirac equation, can be obtained from Eq. (2.13.4) defining $\boldsymbol{B}^{(3)}$ independently from considerations of *O(3)* gauge geometry, through the charge quantization condition (2.13.5) in the form $\boldsymbol{p}^{(1)} = e\boldsymbol{A}^{(1)}$. The theory is therefore consistent.

13.4 Discussion

Since H_{spin} in Eq. (2.13.6) is a Hamiltonian, it is time independent, showing that $\boldsymbol{B}^{(3)}$ is a phase free, time-independent, and observable component of vacuum electrodynamics. Equation (2.13.1) relates it to the plane waves $\boldsymbol{B}^{(1)} = \boldsymbol{B}^{(2)*}$, which are complex conjugates in the basis (1), (2), (3). It follows from the well known minimal prescription,

$$p_\mu \rightarrow p_\mu + eA_\mu \,, \tag{2.13.8}$$

(the basis [18] of the Aharonov-Bohm effect) that the transverse momenta of the electron in equilibrium with the field can be represented in the same basis by the complex conjugate pairs,

$$\boldsymbol{p}^{(1)} = \boldsymbol{p}^{(2)*} \,. \tag{2.13.9}$$

In so doing, it is understood that measurable quantities are real, physical observables, as in electrodynamics in general. The electron transverse momentum is driven by the field transverse momentum in field-electron equilibrium. This requires

$$e\boldsymbol{A}^{(1)} = \boldsymbol{p}^{(1)} = \hbar\boldsymbol{\kappa}^{(1)} = i\hbar\nabla^{(1)} \,, \tag{2.13.10}$$

and taking magnitudes on both sides leads to the charge quantization condition. The electron property (orbital angular momentum) is created from the electromagnetic field, and the charge quantization condition the electron property and the field property are indistinguishable.

Therefore, although the photon is conventionally considered to be uncharged, its quantized momentum $\hbar\kappa$ is now understood to have the classical value $eA^{(0)}$, the product of two $\hat{C}$ negative quantities. At a fundamental level, therefore, the charge on the electron e becomes the *O(3)* gauge coupling parameter, the constant of proportionality between momentum and the vector potential. This is a result of the *O(3)* symmetry itself [6,12], and so in this view, the vector potential is physically meaningful. This is confirmed in the Aharonov-Bohm effect [18] which has deeply meaningful consequences, for example in vacuum topology [12]. These inferences all rest on the emergence of $\boldsymbol{B}^{(3)}$, and illustrate its central importance in field-particle theory.

Acknowledgments

It is a pleasure to acknowledge many interesting discussions with several colleagues, among whom are: Yildirim Aktas, Ahmed Hasanein, the late Stanislaw Kielich, Mikhail Novikov, Mark P. Silverman, Jean-Pierre Vigier, and B. Yu. Zel'dovich.

References

[1] M. W. Evans, *Physica B* **182**, 227, 237 (1992).

[2] M. W. Evans, *Physica B* **183**, 103 (1993); 190, 310 (1993).

[3] M. W. Evans, *The Photon's Magnetic Field.* (World Scientific, Singapore, 1992).

[4] M. W. Evans, and S. Kielich, eds., *Modern Nonlinear Optics,* Vol. 85(2) of *Advances in Chemical Physics,* I. Prigogine and S. A. Rice, eds., (Wiley Interscience, New York, 1993).

[5] M. W. Evans and A. A. Hasanein, *The Photomagneton and Quantum Field Theory.* (World Scientific, Singapore, 1994).
[6] M. W. Evans and J.-P. Vigier, *The Enigmatic Photon, Volume 1: The Field* $\boldsymbol{B}^{(3)}$. (Kluwer, Dordrecht, 1994), *The Enigmatic Photon, Volume 2: Non-Abelian Electrodynamics.* (Kluwer Academic, Dordrecht, 1995).
[7] M. W. Evans, *Mod. Phys. Lett.* **7**, 1247 (1993).
[8] M. W. Evans, *Found. Phys. Lett.* **7**, 67 (1994).
[9] M. W. Evans, *Found. Phys. Lett.* in press (1994); and submitted.
[10] M. W. Evans, *Found. Phys.* in press (1995).
[11] J. Deschamps, M. Fitaire, and M. Lagoutte, *Rev. Appl. Phys.* **7**, 155 (1972).
[12] L. H. Ryder, *Quantum Field Theory*, 2nd edn. (Cambridge University Press, Cambridge, 1987).
[13] G. H. Wagnière, *Linear and Nonlinear Optical Properties of Molecules.* (Verlag Helvetica Chimica Acta, Basel, 1993).
[14] R. Zawodny, Ref. 4, Vol. 85(1), a review; J. P. van der Ziel, P. S. Pershan, and L. D. Malmstrom, *Phys. Rev.* **143**, 574 (1966).
[15] N. Sanford, R. W. Davies, A. Lempicki, A. J. Miniscalco, and S. J. Nettel, *Phys. Rev. Lett.* **50**, 1803 (1983).
[16] W. Happer, *Rev. Mod. Phys.* **44**, 169 (1972).
[17] J. Deschamps, M. Fitaire, and M. Lagoutte, *Phys. Rev. Lett.* **25**, 1330 (1970).
[18] M. P. Silverman, *Phys. Lett. A* **182**, 323 (1993).

Paper 14

The Evans-Vigier Field, $\boldsymbol{B}^{(3)}$, in Dirac's Original Electron Theory: a New Theorem of Field-Fermion Interaction

Dirac's original electron theory is used to show that a classical electromagnetic field interacts with quantized fermion half integral spin through the Evans-Vigier field, $\boldsymbol{B}^{(3)} = -i(e/\hbar)\boldsymbol{A} \times \boldsymbol{A}^*$, where $\boldsymbol{A} \times \boldsymbol{A}^*$ is the conjugate product of field vector potential, $\boldsymbol{A}$, with its own complex conjugate $\boldsymbol{A}^*$; and where $e/\hbar$ is the ratio of elementary charge to Dirac constant. Dirac's theory of the electron is recovered when $\boldsymbol{A}^*$ is replaced by $\boldsymbol{A}$. However, since $\boldsymbol{A}$ is complex from d'Alembert's equation in vacuo, $\boldsymbol{B}^{(3)}$ is always non-zero. It becomes very large at low frequencies for moderate field intensity, and has several important practical applications.

14.1 Introduction

The original description by Dirac [1] of his famous theory of the electron is used in this communication to show that the classical electromagnetic field interacts with quantized fermion spin through the Evans-Vigier field [2—10],

$$\boldsymbol{B}^{(3)} = -i\frac{e}{\hbar}\boldsymbol{A}\times\boldsymbol{A}^{*}, \tag{2.14.1}$$

where $\boldsymbol{A}\times\boldsymbol{A}^{*}$ is the conjugate product of the field vector potential with its own complex conjugate $\boldsymbol{A}^{*}$. Here e is elementary charge and $\hbar$ the Dirac constant, and $\boldsymbol{A}\times\boldsymbol{A}^{*}$ is pure imaginary, so that $\boldsymbol{B}^{(3)}$ is real and physical. Equation (2.14.1) represents a new fundamental theorem in field theory, and can be generalized within quantum electrodynamics [11—14]. The demonstration of Eq. (2.14.1) is given in Sec 14.2, and is based closely on the original description by Dirac [1]. The latter's theory is recovered exactly if $\boldsymbol{A}^{*}$ in Eq. (2.14.1) and related equations is replaced by $\boldsymbol{A}$. In other words, Dirac assumes [1] that $\boldsymbol{A}$ is pure real, so that $\boldsymbol{A}\times\boldsymbol{A}^{*}$ (and $\boldsymbol{B}^{(3)}$) is zero in his theory. More generally however, the d'Alembert equation in vacuo [11—14] shows that $\boldsymbol{A}$ is complex, with a real and imaginary part. In consequence the cross product [2—10] $\boldsymbol{A}\times\boldsymbol{A}^{*}$ is not zero. For a transverse plane wave $\boldsymbol{A}$ the conjugate product $\boldsymbol{A}\times\boldsymbol{A}^{*}$ is pure imaginary and free of the electromagnetic phase. It is, furthermore, directly proportional to beam power density I (W m^{-2}) and inversely proportional to the square of beam angular frequency (ω). At low frequencies $\boldsymbol{B}^{(3)}$ becomes very large (megatesla) for moderate I (of the order ten watts per square centimeter). This property is potentially of great practical utility, the existence of $\boldsymbol{B}^{(3)}$ being the result of Dirac's fundamental theory of the electron. In Sec. 14.3 we indicate avenues of generalization of this result within contemporary quantum electrodynamics.

14.2 The Conjugate Product in Dirac's Original Electron Theory

Dirac has given a clear description of his own theory of the electron interacting with a classical electromagnetic field in chapter eleven of Ref. 1. In this section Dirac's description is followed closely to show the existence of the field $\boldsymbol{B}^{(3)}$ of Eq. (2.14.1) for a complex electromagnetic potential four-vector, A_μ, the general solution of the vacuum d'Alembert equation

[15]. The space part of A_μ is denoted $\boldsymbol{A}$, and its complex conjugate by $\boldsymbol{A}^*$. In his original development [1], Dirac assumed that $\boldsymbol{A}$ is classical and real, so that $\boldsymbol{A} \times \boldsymbol{A}^*$ is zero, because he was aiming at a theory of the anomalous Zeeman effect in a *static* magnetic field. Empirical evidence is now available [16—20], however, to show that for electromagnetic waves, $\boldsymbol{A} \times \boldsymbol{A}^*$ is experimentally observable in magneto-optical phenomena, and is non-zero experimentally.

Dirac's development is based [1] on the quantum relativistic wave equation,

$$\left(p_0 + eA_0 - \rho_1(\sigma \cdot (\boldsymbol{p} + e\boldsymbol{A})) - \rho_3 mc\right)\psi = 0, \tag{2.14.2}$$

for an electron (or more generally a fermion) in a classical electromagnetic field. Equation (2.14.2) is written here in contemporary standard *(S.I.)* Units, (whereas Dirac uses Gaussian units). The energy momentum four-vector in *S.I.* units in Dirac's notation is

$$p_\mu := (p_0, \boldsymbol{p}), \tag{2.14.3}$$

and the potential four-vector is

$$A_\mu := (A_0, \boldsymbol{A}). \tag{2.14.4}$$

The matrices ρ_1, and ρ_3 are [1]

$$\rho_1 = \begin{bmatrix} 0 & 0 & 1 & 0 \\ 0 & 0 & 0 & 1 \\ 1 & 0 & 0 & 0 \\ 0 & 1 & 0 & 0 \end{bmatrix}, \quad \rho_3 = \begin{bmatrix} 1 & 0 & 0 & 0 \\ 0 & 1 & 0 & 0 \\ 0 & 0 & -1 & 0 \\ 0 & 0 & 0 & -1 \end{bmatrix}, \tag{2.14.5}$$

and ψ is a column four-vector, described in contemporary terms as the Dirac four-spinor. (In his original account [1] Dirac does not use parity inversion to interrelate spinor components, as is the contemporary practice [11—14].

The classical Hamiltonian for an electron in a classical electromagnetic field is now used by Dirac [1] as a guideline to the properties of Eq. (2.14.2). We proceed here by following this method closely, but by indicating at each stage the modifications which enter into the Dirac theory of the electron when $\boldsymbol{A}$ is complex rather than real. The wave equation expected from analogy with the classical theory is Eq. (2.14.30) of chapter eleven of Dirac [1],

$$\left((p_0 + eA_0)^2 - (\boldsymbol{p} + e\boldsymbol{A})^2 - m^2c^2\right)\psi = 0 , \tag{2.14.6}$$

and is written for real $\boldsymbol{A}$. For complex A_μ Eq. (2.14.6) becomes

$$\begin{gathered}\Big((p_0 + eA_0)(p_0 + eA_0^*) - \\ (\boldsymbol{p} + e\boldsymbol{A})\cdot(\boldsymbol{p} + e\boldsymbol{A}^*) - m^2c^2\Big)\psi = 0 .\end{gathered} \tag{2.14.7}$$

In order to make his theory of the electron resemble Eq. (2.14.2) as closely as possible, Dirac multiplies Eq. (2.14.2) by the factor $p_0 + eA_0 + \rho_1(\sigma\cdot(\boldsymbol{p} + e\boldsymbol{A})) + \rho_3 mc$, which for a complex potential four-vector becomes $p_0 + eA_0^* + \rho_1(\sigma\cdot(\boldsymbol{p} + e\boldsymbol{A}^*)) + \rho_3 mc$, giving the product

$$\Big((p_0 + eA_0^*)(p_0 + eA_0) - (\sigma\cdot(\boldsymbol{p} + e\boldsymbol{A}^*)) \times (\sigma\cdot(\boldsymbol{p} + e\boldsymbol{A})) - m^2c^2 - \rho_1\Big((p_0 + eA_0^*)(\sigma\cdot(\boldsymbol{p} + e\boldsymbol{A})) - (\sigma\cdot(\boldsymbol{p} + e\boldsymbol{A}^*))(p_0 + eA_0)\Big)\Big)\psi = 0 . \tag{2.14.8}$$

This replaces Eq. (2.14.31) of Dirac's original theory [1]. The product (2.14.8) contains several terms which are developed as follows. The conjugate product term leading to Eq. (2.14.1) originates in $e^2(\sigma\cdot\boldsymbol{A}^*)(\sigma\cdot\boldsymbol{A})\psi$. As shown by Dirac [1], if $\boldsymbol{B}$ and $\boldsymbol{C}$ are any two three-dimensional vectors that commute with σ, then

$$(\sigma\cdot\boldsymbol{B})(\sigma\cdot\boldsymbol{C}) = \boldsymbol{B}\cdot\boldsymbol{C} + i(\sigma\cdot\boldsymbol{B}\times\boldsymbol{C}) . \tag{2.14.9}$$

In contemporary terms σ is known as the Pauli matrix [11—14]. For a pure real $\boldsymbol{A}$, there is only one term on the right hand side of Eq. (2.14.9), but for a complex $\boldsymbol{A}$, there enters into the Dirac theory of the electron a new term, which describes the interaction of the conjugate product $\boldsymbol{A}\times\boldsymbol{A}^*$ of the classical electromagnetic field with the non-classical matrix σ. This appears to be an exceedingly useful new result, because resonance can be induced between the two spin states of the fermion, in direct analogy with *NMR* or *ESR*, but instead of using a cumbersome permanent magnet (static magnetic field) to do this we may now use an ordinary radio frequency electromagnetic field generator.

Following the development by Dirac [1], but allowing now for complex $\boldsymbol{A}$, we set

$$\boldsymbol{p} \rightarrow -i\hbar\nabla, \tag{2.14.10}$$

and obtain the terms

$$1)\ \boldsymbol{p} \times \boldsymbol{p} = \mathbf{0}, \tag{2.14.11a}$$

$$2)\ e\boldsymbol{A} \times \boldsymbol{p} = \mathbf{0}, \tag{2.14.11b}$$

$$3)\ e\boldsymbol{p} \times \boldsymbol{A}^* = -i\hbar e\nabla \times \boldsymbol{A}^* = -i\hbar e\boldsymbol{B}^*. \tag{2.14.11c}$$

The last of these uses the quantum prescription Eq. (2.14.10) to describe the interaction of the intrinsic spin of the fermion with the magnetic component of the electromagnetic field,

$$\boldsymbol{B}^* = \nabla \times \boldsymbol{A}^*, \tag{2.14.12}$$

and for a static magnetic field, leads to the famous result [1] that the intrinsic spin angular momentum of a fermion is half integral in the non-relativistic, non-classical limit. In our case, $\boldsymbol{B}^*$ is an electromagnetic plane wave and averages to zero over many cycles of the field. Term three is therefore of no further interest in our analysis. Similarly, we follow Dirac in discarding term one, the cross product $\boldsymbol{p} \times \boldsymbol{p}$. The term $e\boldsymbol{A} \times \boldsymbol{p}$ is also discarded, as usual, because it is a classical quantity multiplying a del operator, $\boldsymbol{p} = -i\hbar\nabla$ [1], operating on zero. (In contrast, the term $e\boldsymbol{p} \times \boldsymbol{A}^*$ becomes $-ie\hbar\nabla \times \boldsymbol{A}^*$, which is a del operator on $\boldsymbol{A}^*$, and this is non-zero.)

Continuing in this way we find that Dirac's original equation (2.14.34) of chapter eleven of his classic text [1] is replaced for complex $\boldsymbol{A}$ by

$$\begin{aligned}\Big((p_0 + eA_0^*)(p_0 + eA_0) - (\boldsymbol{p} + e\boldsymbol{A}^*)\cdot(\boldsymbol{p} + e\boldsymbol{A})\\ - ie^2\sigma\cdot\boldsymbol{A}^* \times \boldsymbol{A} - m^2c^2 - e\hbar\sigma\cdot\boldsymbol{B}^* + i\rho_1 e\hbar\sigma\cdot\boldsymbol{E}\Big)\psi = 0,\end{aligned} \tag{2.14.13}$$

in which there appears the conjugate product term, and in which products involving real $\boldsymbol{A}$ are suitably modified using elementary complex algebra (i.e., real, physical, quantities are obtained by multiplying complex ones by their complex conjugates). The term in $\sigma \cdot \boldsymbol{e}$ in our Eq. (2.14.13) can contain a real part if $\boldsymbol{E}$ is a complex plane wave, but the symmetry of this real part is $\hat{T}$ negative, $\hat{P}$ negative [21], and this is not physical. (This is the same kind of reasoning used to argue that there is no electric counterpart of the Faraday effect, or inverse Faraday effect [16—20].) For this reason, and also because $\boldsymbol{E}$ is oscillatory and averages to zero over many field cycles, we take no further interest here in this term. In so doing it is assumed that the term,

$$\text{Real}\ (A_0^* \boldsymbol{A} - \boldsymbol{A}^* A_0) = \boldsymbol{0}. \tag{2.14.14}$$

In the transverse gauge, $A_0 = 0$, and in the Coulomb gauge, A_0 is a real constant which may be zero, so Eq. (2.14.14) is satisfied in both gauges. There may, conceivably, be a gauge in which $A^{(0)}$ is complex, but the physical results of Dirac's electron theory must always be gauge independent.

So the final result of our calculation is

$$\Big((p_0 + eA_0^*)(p_0 + eA_0) - (\boldsymbol{p} + e\boldsymbol{A}^*) \cdot (\boldsymbol{p} + e\boldsymbol{A}) - m^2c^2 \\ - ie^2 \sigma \cdot \boldsymbol{A}^* \times \boldsymbol{A}\Big)\psi = 0, \tag{2.14.15}$$

in which there appears the gauge independent term $-ie^2\sigma \cdot \boldsymbol{A}^* \times \boldsymbol{A}$ which allows resonance to occur between the two energy states of the matrix σ, the conjugate product $\boldsymbol{A}^* \times \boldsymbol{A}$ playing the role of a magnetic field defined by Eq. (2.14.1) [2—10]. This is a result of Dirac's relativistic quantum theory of the electron in a classical electromagnetic field. More rigorously, it can also be obtained with contemporary quantum electrodynamics [11—14], in which there are small radiative corrections leading to phenomena such as the Lamb shift and the anomalous magnetic moment of the electron [11—14].

The non-relativistic limit of Eq. (2.14.15) can be obtained using the standard approximations [1],

$$En \sim mc^2, \qquad \boldsymbol{p} \sim \boldsymbol{0}, \tag{2.14.16}$$

in which case we obtain the interaction energy eigenvalue [1, 2, 14],

$$W := En - mc^2 = \frac{e^2c^2(\sigma \cdot \boldsymbol{A})(\sigma \cdot \boldsymbol{A}^*)}{En + mc^2 + ecA_0} - ecA_0 . \tag{2.14.17}$$

In the radiation gauge, $A_0 = 0$, and this result reduces to

$$W := En - mc^2 = \frac{e^2}{2m}(\boldsymbol{A} \cdot \boldsymbol{A}^* + i\sigma \cdot \boldsymbol{A} \times \boldsymbol{A}^*) . \tag{2.14.18}$$

Using Eq. (2.14.1), the resonance term becomes the familiar equation for the anomalous Zeeman effect,

$$W_R = -\frac{e}{m}\frac{\hbar}{2}\sigma \cdot \boldsymbol{B}^{(3)}, \tag{2.14.19}$$

a result which shows that the Evans-Vigier field $\boldsymbol{B}^{(3)}$ always exists in field-fermion interaction. In other words, the classical electromagnetic field acts on the fundamental half integral spin of a fermion as if it were a magnetic field, $\boldsymbol{B}^{(3)}$. We shall see in Sec. 14.3 that this field has very useful properties.

14.3 Properties of $\boldsymbol{B}^{(3)}$ in Classical and Quantum Electrodynamics

In terms of intensity (I) and angular frequency (ω), $\boldsymbol{B}^{(3)}$ is [22,23],

$$\boldsymbol{B}^{(3)} = \frac{e\mu_0 c}{\hbar}\frac{I}{\omega^2}\boldsymbol{e}^{(3)} = 5.723 \times 10^{16}\frac{I}{\omega^2}\boldsymbol{e}^{(3)}, \tag{2.14.20}$$

where $\boldsymbol{e}^{(3)}$ is a unit vector in the propagation axis (3) of the radiation, and μ_0 the permeability in vacuo in *S.I.* units [24]. For a given intensity, therefore, $\boldsymbol{B}^{(3)}$ is inversely proportional to the square of field angular frequency (radians $s^{-1} = 2\pi$ f where f is in hertz, or cycles per second). Whenever the classical electromagnetic field interacts with a fermion of quantum mechanics the $\boldsymbol{B}^{(3)}$ field generates a resonance effect between the two states of the non-classical spinor. The resonance occurs, as usual [25—28], when a photon of probe radiation, $\hbar\omega_{res}$, is absorbed to induce a change between the lower and upper energy states defined by the mathematical properties of the spinor, in this case the third Pauli matrix [1, 2, 14],

$$\sigma_Z = \sigma^{(3)} = \begin{pmatrix} 1 & 0 \\ 0 & -1 \end{pmatrix}. \tag{2.14.21}$$

In contemporary terms this is referred to as a *spin flip* between the half integral spin (angular momentum) states of the fermion. In *NMR* the fermion may be a proton, or a neutron, in *ESR* an electron. The factor two which gives rise to the everyday term *half integral fermion spin* is a consequence, however, of an approximation, (demonstrated in Sec. 14.2), and the fundamental reason for the existence of *NMR* and *ESR* can be traced to topology [11—14], in that the group space of a fermion is different from that of a boson. Thus *NMR* and *ESR* are examples of absorption spectroscopies [29] based on the Dirac equation, which is solved in a non-relativistic limit. It is advantageous to bear in mind that Dirac derived the equation purely from the general principles of quantum mechanics and special relativity [1]. These considerations (e.g. that the wave equation must be linear in p_0 and $\boldsymbol{p}$) force the use of anti-commuting 4 x 4 matrices, of which the Pauli matrix are component 2 x 2 matrices. Therefore the fermion intrinsic spin has a deeper meaning than angular momentum, and it is well known that the fermion spin cannot be pictured classically (e.g. as a spinning object in space).

There is no reason, therefore, to assume that *NMR* and /or *ESR* must always be practiced with static magnetic fields, (or that a Pauli matrix must always interact with a static magnetic field) and Sec. 14.2 has shown that the conjugate product $\boldsymbol{A} \times \boldsymbol{A}^*$ is a result of the Dirac equation of a fermion in a classical electromagnetic field, using only the standard *minimal prescription* [11—14],

$$p_\mu \rightarrow p_\mu + eA_\mu. \tag{2.14.22}$$

This prescription of relativistic quantum field theory [11—14] is well known to be the result of type two (local) gauge invariance, which is a fundamental assumption in contemporary orthodoxy. It is also well established that the conjugate product produces various observable magneto-optic effects, prominent among which is the inverse Faraday effect [16—18]. Therefore Sec. 14.2 shows (as far as we are aware, for the first time) that the Dirac equation produces the inverse Faraday effect. This is a reassuring result both for field theory and experimental magneto-optics.

For our purposes the Evans-Vigier field, $\boldsymbol{B}^{(3)}$, from Eq. (2.14.1) is orders of magnitude more intense for a given I at radio frequencies (MHz) rather than at visible frequencies (100 to 1000 THz). This is simply the result of its inverse square dependence on field frequency. For I, for example, of 10.0 watts per square centimeter, the $\boldsymbol{B}^{(3)}$ field reaches an order of magnitude of nanotesla at 5,000 cm^{-1} in the visible, (a hundred thousand times weaker than the Earth's mean magnetic field), but for a 10.0 *MHz* radio frequency field it becomes 1.45 *mega*tesla, causing proton resonance in the *infra-red* at about two thousand wavenumbers [22,23]. This proton resonance frequency (a spectral absorption feature) is of course observable with infra red radiation in a contemporary spectrometer [29]. (It is actually much easier to observe in the infra red, because the need for the resonance causing, rotating, radio frequency field [25—28] in ordinary *NMR* spectrometers is removed.) These simple calculations therefore reveal an *enormous* potential resolution advantage over contemporary *NMR* (of all varieties) because the latter is practiced with a static magnetic field [25—28] of the order 10 tesla maximum. Much of the effort in contemporary *NMR* practice [25—28] revolves around the (expensive) technology of superconducting magnets, whose fields reach, perhaps, 25 tesla [25—28], with great effort and ingenuity. (World records for these fields are claimed regularly.) This results in proton resonance at, say 0.5 *GHz*, around which the chemical shift structure is observed as fine detail, in one, two, or three correlation dimensions [25—28] with many more or less exotic variations in pulse sequences. With the use of $\boldsymbol{A} \times \boldsymbol{A}^*$, and ordinary radio frequency generators, it appears perfectly feasible to advance this 0.5 *GHz* resonance frequency into the infra red (*THz* range) as just described, making superconducting magnets unnecessary. The advantages of such a technology, if realized, are bounded only by the imagination and art of the spectroscopist.

The theory in Sec. 14.2 is based on a classical electromagnetic field, whereas more rigorously, there are radiative corrections due to quantum electrodynamics (*QED*) [11—14], in which there is an extensive late twentieth century literature. In respect of electron resonance, *QED* leads, as is well known [11—14], to a 1% correction to the factor 2 in Eq. (2.14.19). Therefore in practical *NMR* and *ESR*, *QED* does not play a central role. In the delicate interplay between electron and photon however, *QED* is all-important, and future developments in $\boldsymbol{B}^{(3)}$ theory should aim to quantize the electromagnetic field as it interacts with the already quantized fermion. The anomalous Landé factor of the electron, first discussed by Schwinger [30], should in theory become observable with $\boldsymbol{A} \times \boldsymbol{A}^*$ rather than with a static magnetic field, and the original experimental measurements, ably described again by Dirac [1,31], should be repeatable with an optically generated $\boldsymbol{A} \times \boldsymbol{A}^*$. This line of reasoning can clearly be extended to all magnetic effects, of which there are many now known. In each case, the static magnetic field is replaced by the Evans-Vigier field,

$\boldsymbol{B}^{(3)}$, of the incoming electromagnetic beam, or photon beam. There should probably be non-classical photon statistical effects akin to light squeezing and a whole variety of new optical resonance phenomena should eventually emerge now that the existence of $\boldsymbol{A} \times \boldsymbol{A}^*$ is proven from the Dirac equation.

Acknowledgments

York University, Toronto, Department of Physics and Astronomy, is thanked for an invitation to spend a year there as visiting professor. Prof. Stanley Jeffers and Sisir Roy are thanked for many interesting discussions.

References

[1] P. A. M. Dirac, *Quantum Mechanics*, 4th edn., Chap.11, (Oxford University Press, Oxford, 1974).

[2] M. W. Evans and J.-P. Vigier, *The Enigmatic Photon, Volume 1: The Field* $\boldsymbol{B}^{(3)}$ (Kluwer Academic Publishers, Dordrecht, 1994); ibid., *The Enigmatic Photon, Volume 2: Non-Abelian Electrodynamics* (Kluwer Academic Publishers, Dordrecht, 1995).

[3] A. A. Hasanein and M. W. Evans, *The Photomagneton in Quantum Field Theory* (World Scientific, Singapore, 1994).

[4] M. W. Evans and S. Kielich, eds., *Modern Nonlinear Optics,* Vols. 85(1), 85(2), 85(3) of *Advances in Chemical Physics,* I. Prigogine and S. A. Rice, eds., (Wiley Interscience, New York, 1993).

[5] M. W. Evans, *The Photon's Magnetic Field* (World Scientific, Singapore, 1992).

[6] M. W. Evans in A. van der Merwe and A. Garuccio, eds., *Waves and Particles in Light and Matter* (Plenum, New York, 1994), the de Broglie centennial conference volume.

[7] M. W. Evans, *Physica B* **182**, 227, 237 (1992); **183**, 103 (1993); **190**, 310 (1993); *Physica A*, in press (1995); *Physica A*, submitted for publication.

[8] M. W. Evans and S. Roy, *Physica A*, submitted for publication; S. Roy and M. W. Evans, *Found. Phys.*, submitted for publication.
[9] M. W. Evans, *Found. Phys. Lett.* **7**, 67, 209, 379, 467, 577 (1994); *Found. Phys.* **24**, 1519, 1671 (1994); **25**, 175, 383 (1995).
[10] O. Costa de Beauregard, *Found. Phys. Lett.*, in press (1995).
[11] L. H. Ryder, *Quantum Field Theory*, 2nd. edn. (Cambridge University Press, Cambridge, 1987).
[12] C. Itzykson and J.-B. Zuber, *Quantum Field Theory* (McGraw-Hill Intermational, New York, 1980).
[13] J. D. Bjorken and S. D. Drell, *Relativistic Quantum Mechanics* (McGraw-Hill International, New York, 1964).
[14] A. Messiah, *Quantum Mechanics*, Vol. 2 (North Holland, Amsterdam, 1962).
[15] J. D. Jackson, *Classical Electrodynamics* (Wiley, New York, 1962); L. D. Landau and E. M. Lifshitz, *The Classical Theory of Fields*, 4th. edn. (Pergamon, Oxford, 1975).
[16] R. Zawodny, in Vol. 85(1) of Ref. 4, a review of magneto-optics.
[17] G. H. Wagnière, *Linear and Nonlinear Optical Properties of Molecules* (VCH, Basel, 1993); G. H. Wagnière, *Phys. Rev. A* **40**, 2437 (1989).
[18] J. P. van der Ziel, P. S. Pershan, and L. D. Malmstrom, *Phys. Rev.* **143**, 574 (1966).
[19] S. Woźniak, M. W. Evans and G. H. Wagnière, *Mol. Phys.* **75**, 81, 99 (1992).
[20] T. W. Barrett, H. Wohltjen and A. Snow, *Nature* **301**, 694 (1983).
[21] M. W. Evans, in I. Prigogine and S. A. Rice, eds., *Advances in Chemical Physics*, Vol. 81, (Wiley-Interscience, New York, 1992).
[22] M. W. Evans, York University Internal Report 1 (1995).
[23] M. W. Evans, York University Internal Report 3 (1995).
[24] P. W. Atkins, *Molecular Quantum Mechanics, 2nd. edn.* (Oxford University Press, Oxford, 1983).
[25] J. W. Hennel and J. Klinowsky, *Fundamentals of NMR* (Longmuir, London, 1993).
[26] C. P. Slichter, *Principles of Magnetic Resonance, 2nd. edn.* (Springer, Berlin, 1978).

[27] S. W. Homans, *A Dictionary of Concepts in NMR* (Clarendon, Oxford, 1989).

[28] R. R. Ernst, G. Bodenhausen and A. Wokaun, *Principles of Nuclear Magnetic Resonance in One and Two Dimensions* (Oxford University Press, Oxford, 1987).

[29] M. W. Evans, G. J. Evans, W. T. Coffey, and P. Grigolini, *Molecular Dynamics* (Wiley-Interscience, New York, 1982).

[30] described in W. Heitler, *The Quantum theory of Radiation*, 3rd. edn. (Oxford University Press, Oxford, 1954).

[31] P. A. M. Dirac, *Lectures on Quantum Field Theory* (Belfer Graduate School, Yeshiva University, New York, 1966).

Paper 15

The Microwave Optical Zeeman Effect Due to $\boldsymbol{B}^{(3)}$

The optical conjugate product of a circularly polarized laser is used in the Dirac equation to show the presence of a microwave frequency optical Zeeman effect which is proportional at a given angular frequency to the Evans-Vigier field $\boldsymbol{B}^{(3)}$ of the microwave radiation. An experimental arrangement to detect this effect is proposed, using ESR technique.

Key words: Microwave optical Zeeman effect, $\boldsymbol{B}^{(3)}$ field.

15.1 Introduction

Recently, it has been demonstrated that the Dirac equation of one electron in a circularly polarized electromagnetic field can be solved to show the existence of the $\boldsymbol{B}^{(3)}$ (Evans-Vigier) field, a magnetic flux density whose classical source is the conjugate product $\boldsymbol{B}^{(1)} \times \boldsymbol{B}^{(2)}$ of plane wave solutions of Maxwell's equations in the vacuum. In the appropriate circular basis [1—5] there exist the cyclically symmetric relations between fields,

$$\boldsymbol{B}^{(1)} \times \boldsymbol{B}^{(2)} = iB^{(0)}\boldsymbol{B}^{(3)*}, \text{ et cyclicum,} \tag{2.15.1}$$

so that $\boldsymbol{B}^{(3)}$ is phase free. Here $B^{(0)}$ is a scalar amplitude (tesla). Whenever radiation magnetizes matter, the effect depends on $\boldsymbol{B}^{(3)}$ at first and second order [6]. At visible frequencies, there is an inverse Faraday effect and optical Zeeman effect which are linear [7] in the beam power density, I, (in watts per square meter). These effects originate in $iB^{(0)}\boldsymbol{B}^{(3)*}$. At microwave frequencies and sufficient power densities, however, the inverse Faraday effect becomes proportional directly [8—10] to the beam's $\boldsymbol{B}^{(3)}$ field, and thus to $I^{1/2}$. Since $\boldsymbol{B}^{(3)}$ travels at the speed of light in vacuo, and cannot exist in isolation of its source, the conjugate product $\boldsymbol{B}^{(1)} \times \boldsymbol{B}^{(2)}$, there is no free Faraday induction, i.e., a modulated laser sent through an induction coil without a sample will produce no signal in the coil. The reason for this is that $\boldsymbol{B}^{(3)}$ travels at c in the vacuum and under these conditions the only electric fields allowed by symmetry and relativity are the ordinary, transverse, plane waves $\boldsymbol{E}^{(1)}$ and $\boldsymbol{E}^{(2)}$, these being complex conjugate pairs in the basis (1), (2), (3).

In this Letter, the Dirac equation is used to show that there exists a microwave frequency optical Zeeman effect which for a given microwave pump frequency ω is directly proportional to the $\boldsymbol{B}^{(3)}$ field of the radiation. In Sec. 15.2, the Dirac equation is solved for the interaction of the beam conjugate product with one electron. At visible frequencies this produces the optical Zeeman effect, which is proportional to the beam power density I. At microwave frequencies the Dirac equation produces an optically induced Zeeman effect proportional to the square root of I. At these frequencies the beam property producing the effect is $\boldsymbol{B}^{(3)}$. Section 15.3 is a discussion of this result in terms of transfer of photon angular energy to the electron, and it is shown that the result of the calculation in Sec. 15.2 is consistent with conservation of angular energy in a photon-electron collision.

15.2 The Optical Zeeman Effect from the Dirac Equation

Since $\boldsymbol{B}^{(3)}$ cannot exist in isolation of its source, the electromagnetic conjugate product (Eq.(2.15.1)), we calculate the optical Zeeman effect from the pure electromagnetic term in the Dirac equation [6] of an electron in a circularly polarized electromagnetic beam,

$$\hat{W}u = \left(\frac{ie^2}{2m_0 + eA^{(0)}/c} \sigma \cdot \boldsymbol{A}^{(1)} \times \boldsymbol{A}^{(2)} \right) u. \qquad (2.15.2)$$

Here u is a Dirac four-spinor in the standard representation [6], $\hat{W}$ is an energy eigenoperator whose eigenvalue is given within brackets on the right hand side. In Eq. (2.15.2) e is the charge on the electron, σ a Pauli matrix, m_0 the electron mass, and c the speed of light in vacuo. The conjugate product is expressed as $\boldsymbol{A}^{(1)} \times \boldsymbol{A}^{(2)}$, where $\boldsymbol{A}^{(1)}$ is a vector potential plane wave [6] and $\boldsymbol{A}^{(2)}$ its complex conjugate. The scalar amplitude of $\boldsymbol{A}^{(1)}$ is $A^{(0)}$, and the minimal prescription [6] has been used to describe the momentum and energy imparted relativistically to the electron by the field.

Equation(2.15.2) contains no reference to any field free electron momentum, and uses the rest frame approximation [6] $En \sim m_0 c^2$ for the electron energy. Non-relativistically, therefore, there would be no electron energy in the absence of the beam. The equation also assumes that the scalar potential is $A^{(0)}/c$, and not zero as in the Coulomb gauge. This means physically that the beam imparts energy to the electron as well as momentum. Such a picture is compatible with the Lorentz gauge [6] and a manifestly covariant A_μ four-vector.

From Eq.(2.15.2), the Hamiltonian expectation value (i.e., the energy eigenvalue) is,

$$\langle H \rangle = \frac{ie^2 c}{2m_0 c + eA^{(0)}} \sigma \cdot \boldsymbol{A}^{(1)} \times \boldsymbol{A}^{(2)}, \qquad (2.15.3)$$

which is the relativistic expression of the optical Zeeman effect. Here σ is a Pauli matrix, so we are dealing with the relativistic half-integral spin of the electron in a classically expressed field. The optical Zeeman effect at visible and microwave frequencies emerges by a simple consideration of limits as follows.

(1) When the electron rest momentum is much greater than that imparted by the beam to the electron,

$$2m_0 c \gg eA^{(0)}, \tag{2.15.4}$$

the energy eigenvalue becomes,

$$<H> \rightarrow i\frac{e^2}{2m_0}\sigma \cdot \boldsymbol{A}^{(1)} \times \boldsymbol{A}^{(2)}, \tag{2.15.5}$$

which is to order I, i.e., proportional to the beam power density. Using,

$$A^{(0)} = \frac{B^{(0)}}{\kappa} = \frac{c}{\omega}B^{(0)}, \tag{2.15.6}$$

the condition (4) becomes,

$$\omega \gg \frac{e}{2m_0}B^{(0)}, \tag{2.15.7}$$

which is satisfied at *visible* frequencies for all but enormous, unattainable I. At visible frequencies the Zeeman shift is therefore twice the energy in Eq.(2.15.5), which can be expressed as,

$$<H> \rightarrow -\frac{e^2c^2}{2m_0\omega^2}\sigma \cdot B^{(0)}\boldsymbol{B}^{(3)}. \tag{2.15.8}$$

The term,

$$\chi' := -\frac{e^2c^2}{2m_0\omega^2}, \tag{2.15.8a}$$

is the one electron susceptibility [6].

(2) In the opposite limit, when the momentum imparted by the beam to the electron is much greater than the electron rest momentum,

$$eA^{(0)} \gg 2m_0c, \tag{2.15.9}$$

Eq.(2.15.3) reduces to,

$$\langle H\rangle \rightarrow i\frac{ec}{A^{(0)}}\sigma\cdot\boldsymbol{A}^{(1)}\times\boldsymbol{A}^{(2)}. \tag{2.15.10}$$

The limit(2.15.9) can be rewritten as

$$\omega \ll \frac{e}{2m_0}B^{(0)}, \tag{2.15.11}$$

which is attainable with *microwave* pulses of high power density [11]. Using in Eq.(2.15.10) the relation [6],

$$\boldsymbol{A}^{(1)}\times\boldsymbol{A}^{(2)} = iA^{(0)2}\boldsymbol{e}^{(3)}, \tag{2.15.12}$$

the energy becomes,

$$\langle H\rangle \underset{m_0\rightarrow 0}{\rightarrow} -ec\sigma\cdot A^{(0)}\boldsymbol{e}^{(3)} = -\frac{ec^2}{\omega}\sigma\cdot\boldsymbol{B}^{(3)}, \tag{2.15.13}$$

which for a given angular frequency, ω, is proportional to $\boldsymbol{B}^{(3)}$ of the beam and therefore [11] to the square root of its power density. Equation(2.15.13) shows that the optical Zeeman effect at microwave frequencies is determined entirely by two beam properties, ω and $\boldsymbol{B}^{(3)}$.

15.3 Discussion

Equation(2.15.13) can be rewritten as

$$<H> \underset{m_0 \to 0}{\to} -ceA^{(0)}\sigma \cdot \boldsymbol{e}^{(3)}, \tag{2.15.14}$$

which is seen to have the correct units of energy (because $\sigma \cdot \boldsymbol{e}^{(3)}$ is unitless) and because $eA^{(0)}$ is electron momentum magnitude acquired from the beam.

Using the charge quantization condition [6,12], which is implied by the existence of $\boldsymbol{B}^{(3)}$ [12],

$$eA^{(0)} = \hbar\kappa, \tag{2.15.15}$$

Eq.(2.15.14) becomes,

$$<H> \underset{m_0 \to 0}{\to} -\hbar\omega\sigma \cdot \boldsymbol{e}^{(3)}, \tag{2.15.16}$$

and this shows that the rotational energy, $\hbar\omega$, of the photon has been transferred completely to the electron. In this limit, the expected Zeeman splitting is therefore $2\hbar\omega$. This limit can never be attained in practice because the rest momentum of the electron is always non-zero in special relativity, but under condition(2.15.11), it can be approximated.

The equation(2.15.2), which starts from the conjugate product of the classical field, has therefore produced the result expected on the grounds of conservation of rotational energy in a perfectly elastic collision between a photon and an electron. This result was attained through the condition(2.15.11), and is consistent with the fact that the photon has lost $\hbar\omega$ and the electron has gained $\hbar\omega$ in this limit. The electron has absorbed the photon and acquired the photon spin. Since the electron has spin states determined by the matrix σ, the spin up state acquires energy $\hbar\omega$, and the spin down state acquires $-\hbar\omega$ for the same sense of circular polarization in the beam (i.e., left or right).

It is to be noted that if $\boldsymbol{B}^{(3)}$ =? $\mathbf{0}$ then this photon absorption process cannot occur at microwave or at visible frequencies. Thus $\boldsymbol{B}^{(3)}$ is a fundamental property of the beam which cannot exist, however, in isolation of its source, the beam conjugate product, which was the optical property used as the starting point of our calculation in Eq.(2.15.2).

It should be possible to see the Zeeman splitting $2\hbar\omega$ by tuning a microwave probe beam to about the frequency 2ω, using ESR technique [13,14]. One configuration which can be suggested is to send an electron or atomic beam in the Z axis, a microwave pump pulse at frequency ω in Y and a microwave probe pulse at about 2ω synchronized with the pulse. The probe should be tunable around 2ω because this is an ideal condition as just discussed.

Acknowledgments

This paper emerged from discussions at a seminar given at York University, Toronto, Canada with several colleagues of the Department of Physics there, and following an invitation by Prof. Stanley Jeffers. These discussions are much appreciated.

References

[1] M. W. Evans, *Physica B* **182**, 227, 237 (1992); **183**, 103 (1993); **190**, 310 (1993).

[2] M. W. Evans, *The Photon's Magnetic Field.* (World Scientific, Singapore, 1992).
[3] M. W. Evans, and S. Kielich, eds., *Modern Nonlinear Optics*, Vols. 85(1), 85(2), 85(3) of *Advances in Chemical Physics,* I. Prigogine and S. A. Rice, eds., (Wiley Interscience, New York, 1993).
[4] M. W. Evans in A. van der Merwe and A. Garuccio, eds., *Waves and Particles in Light and Matter.* (Plenum, New York, 1994), the de Broglie Centennial.
[5] A. A. Hasanein and M. W. Evans, *Quantum Chemistry.*, vol. 1, *The Photomagneton in Quantum Field Theory.* (World Scientific, Singapore, 1994).
[6] M. W. Evans and J.-P. Vigier, *The Enigmatic Photon, Volume 1: The Field* $\boldsymbol{B}^{(3)}$, (Kluwer Academic Publishers, Dordrecht, 1994); ibid., *The Enigmatic Photon, Volume 2: Non-Abelian Electrodynamics*, (Kluwer Academic Publishers, Dordrecht, 1995); *The Enigmatic Photon, Volume 3, A Symposium in Honour of Jean-Pierre Vigier*, conference volume, August 28th to 30th, 1995, at York University, Toronto, in preparation.
[7] P. S. Pershan, *Phys. Rev.* **130**, 919 (1963); J. P. van der Ziel, P. S. Pershan, and L. D. Malmstrom, *Phys. Rev.* **143**, 574 (1966); J. Deschamps, M. Fitaire, and M. Lagoutte, *Phys. Rev. Lett.* **25**, 1330 (1970).
[8] M. W. Evans, *Found. Phys.* **24**, 892, 1519, 1671 (1994).
[9] M. W. Evans, *Found. Phys. Lett.* **7**, 67, 209 (1994).
[10] M. W. Evans, ibid., in press.
[11] S. Roy, *Phys. Lett. A* in press; S. Jeffers, S. Roy and M. W. Evans, *Phys. Lett. A* in prep.; O. Costa de Beauregard, *Found. Phys. Lett.*, in press.
[12] M. W. Evans, *Found. Phys. Lett.*, in press.
[13] P. W. Atkins, *Molecular Quantum Mechanics,* 2nd. edn. (Oxford, 1982).
[14] K. A. Earle, Cornell University, Department of Chemistry, communications.

Paper 16

$\boldsymbol{B}^{(3)}$ Echoes

It is shown that the $\boldsymbol{B}^{(3)}$ field of vacuum electromagnetism regenerates itself throughout spacetime from repeated gauge transforms. These $\boldsymbol{B}^{(3)}$ *echoes* are physical magnetic fields which can be detected experimentally in principle through optical analogues of the Aharonov-Bohm effect.

Key words. Optical Aharonov-Bohm effect, action at a distance, $\boldsymbol{B}^{(3)}$ field

16.1 Introduction

The existence of the $\boldsymbol{B}^{(3)}$ field is established [1—12] by that of magneto-optic effects typified by the well-verified [13—21] inverse Faraday effect. In this note it is argued that the field is echoed throughout space-time by repeated gauge transformations into the vacuum of $\boldsymbol{A}$, where,

$$\boldsymbol{B}^{(3)} := \nabla \times \boldsymbol{A} = -i\frac{e}{\hbar}\boldsymbol{A}^{(1)} \times \boldsymbol{A}^{(2)}, \tag{2.16.1}$$

defines the original $\boldsymbol{B}^{(3)}$ in vacuo in a local region of space-time. Here $\boldsymbol{A}^{(1)} = \boldsymbol{A}^{(2)*}$ is a plane wave potential, a solution of the d'Alembert wave equation, and complex [1—12]. Thus, the conjugate product $\boldsymbol{A}^{(1)} \times \boldsymbol{A}^{(2)}$ is pure imaginary. In the complex space basis ((1), (2), (3)), $\boldsymbol{B}^{(3)}$ is pure real and observable [1—12]. Here $\hbar/e$ is the elementary fluxon (weber) where $\hbar$ is Dirac's constant and e the charge quantum.

The $\boldsymbol{B}^{(3)}$ echoes, in analogy with the Aharonov-Bohm effects [22—25], are physical observables (magnetic flux densities) in regions of space-time where the original $\boldsymbol{B}^{(3)}$ is zero. They can therefore be observed experimentally by carefully excluding the electromagnetic field from direct contact with the sample (for example electrons). Non simply connected vacuum topology [22—25] then supports the existence of non-local effects which are measurable. It is speculated that $\boldsymbol{B}^{(3)}$ echoes might, if observed, be evidence for action at a distance in electromagnetism.

16.2 The Gauge Transformation

Since $\boldsymbol{B}^{(3)}$ is a physical magnetic field it can always be expressed in Eq. (2.16.1) as the curl of a vector potential $\boldsymbol{A}$. The gauge transformation [26],

$$\boldsymbol{A} \rightarrow \boldsymbol{A} + \nabla\phi, \tag{2.16.2}$$

where ϕ is a flux in weber, leaves $\boldsymbol{B}^{(3)}$ unaffected if defined as the curl of $\boldsymbol{A}$. It is therefore invariant under gauge transformation. Since $\boldsymbol{A}^{(1)} \times \boldsymbol{A}^{(2)}$ is an experimental observable [13—21] it is also invariant under gauge transformation. It follows that,

$$\nabla^{(1)} := -i\frac{e}{\hbar}\boldsymbol{A}^{(1)}, \tag{2.16.3}$$

must be regarded as an operator, and that $\boldsymbol{A}^{(2)}$ must be regarded as a vector potential. Equation (2.16.3) is one of the quantum postulates [1—3], i.e., momentum in quantum mechanics is a del operator within a factor $i\hbar$ [22].

The product $\boldsymbol{A}^{(1)} \times \boldsymbol{A}^{(2)}$ is invariant for example under a gauge transform such as,

$$\boldsymbol{A}^{(2)} \rightarrow \boldsymbol{A}^{(2)} + \nabla^{(1)}\phi, \qquad \nabla^{(1)}\phi := \boldsymbol{A}^{(1)}, \tag{2.16.4}$$

in which the del operator is not changed. The del operator is not changed, of course, under the ordinary gauge transform (2.16.2).

In a local region of the vacuum where both $\boldsymbol{A}$ and $\boldsymbol{B}^{(3)}$ are zero, the potential function $\nabla\phi$ is non-zero in general and causes the Aharonov-Bohm effects [22—25]. These are understood as being due to the fact that the vacuum is structured [22]. Recently, optical equivalents of the Aharonov-Bohm effect have been suggested [1—3] and worked out theoretically. Since $\nabla\phi$ is complex and periodic, its conjugate $(\nabla\phi)^*$ is also non-zero, and so there exists the $\boldsymbol{B}^{(3)}$ echo,

$$\boldsymbol{B}_1^{(3)} = -i\frac{e}{\hbar}(\nabla\phi) \times (\nabla\phi)^*, \tag{2.16.5}$$

in regions of the vacuum where $\boldsymbol{B}^{(3)}$ itself is zero experimentally. If $\boldsymbol{B}^{(3)}$ is a magnetic field, then so is $\boldsymbol{B}_1^{(3)}$, and the latter is real, physical, and therefore observable in principle. The process can be continued by gauge transformation on the first echo $\boldsymbol{B}_1^{(3)}$,

$$\boldsymbol{B}_1^{(3)} := \nabla \times \boldsymbol{A}_1, \qquad \boldsymbol{A}_1 \rightarrow \boldsymbol{A}_1 + \nabla\phi_1, \tag{2.16.6}$$

thus defining the second echo in regions of the vacuum where both $\boldsymbol{B}^{(3)}$ and $\boldsymbol{B}_1^{(3)}$ are zero experimentally. This process, if continued, gives an infinite number of echoes,

$$\boldsymbol{B}_1^{(3)}, \ldots, \boldsymbol{B}_n^{(3)}, \qquad n \rightarrow \infty, \tag{2.16.7}$$

which are supported by non-simply-connected vacuum topology [22—25], and are all present in space-time irrespective of any consideration of signal velocity c.

16.3 Non-locality; Action at a Distance

The concept of non-locality can therefore be explained by gauge transforms of this nature, and such an explanation supports the interpretation of quantum mechanics by Bohm and others [26,27]. Although the original $\boldsymbol{B}^{(3)}$ is unchanged by the gauge transform, the $\boldsymbol{B}^{(3)}$ echo is produced nevertheless in a region of space-time where $\boldsymbol{B}^{(3)}$ is zero (for example outside a fibre or waveguide, Sec. 16.4). Similarly the $\boldsymbol{B}_1^{(3)}$ echo produces the $\boldsymbol{B}_2^{(3)}$ echo and so forth for $n \rightarrow \infty$. Therefore the field $\boldsymbol{B}^{(3)}$ is influential in regions infinitely remote from its original locality. This appears to be the first indication of non-locality in an electromagnetic field component rather than in gauge transformed potentials, as in the original Aharonov-Bohm effect [22—25]. Assuming that the field is non-local in this way, its influence is felt in remote regions of space-time without transmittal by a signal velocity, which for the hypothetically massless photon is c. This may therefore be action at a distance, one of a class of superluminal phenomena [28] in electromagnetism. Interestingly, Chubykalo and Smirnov-Rueda [29] have demonstrated the existence of longitudinal solutions of the Maxwell equations in vacuo which involve superluminal and subluminal exponents from the wave equation. Muñera and Guzmán [30] have shown that the reduction of the Maxwell equations to the d'Alembert equation produces a class of longitudinal solutions in vacuo provided that the scalar potential is phase dependent. The $\boldsymbol{B}^{(3)}$ field is therefore an example of a physical longitudinal solution in vacuo with zero phase, and for this reason has the special property of being proportional to the physically observable conjugate product. Gauge transformation of the latter must therefore take place in such a way as to preserve the physical nature of $\boldsymbol{B}^{(3)}$, and as we have seen, this leads to field non-locality (*echoes*), as opposed to potential non-locality.

16.4 Experimental Investigation

A clear experimental demonstration of non-locality in $\boldsymbol{B}^{(3)}$ can be achieved in principle by observing the inverse Faraday effect in regions where the field is excluded. In order to estimate the magnitude of the effect it is sufficient to use a simple classical demonstration based on the relativistic Hamilton-Jacobi equation [1—3] to show the influence of $\boldsymbol{B}^{(3)}$ on one electron. The quantum equivalent of the effect is based on the Dirac equation.

The original inverse Faraday effect was shown by Talin *et al.* [31] to be explicable in terms of the classical, relativistic Hamilton-Jacobi equation. The theory has been developed in terms of $\boldsymbol{B}^{(3)}$ [1] and shows that the energy of interaction of an electron in a circularly polarized electromagnetic field is,

$$\Delta En = \frac{e^2c^2}{\omega}\left(\frac{B^{(0)}}{(m^2\omega^2 + e^2B^{(0)2})^{1/2}}\right)|\boldsymbol{B}^{(3)}| , \qquad (2.16.8)$$

where $\boldsymbol{B}^{(3)} = B^{(0)}\boldsymbol{e}^{(3)}$. Here ω is the field angular frequency and m the mass of the electron. The electronic properties in the interaction energy are e and m; the field properties are ω, c and $B^{(0)}$, the magnitude of $\boldsymbol{B}^{(3)}$ [1]. The plane wave $\boldsymbol{A}^{(1)}$ is a solution of the vacuum d'Alembert equation. In this case, $B^{(0)} = \kappa A^{(0)} = \omega A^{(0)}/c$ [1—3], where $\kappa = \omega/c$ is the wavenumber in vacuo. Equation (2.16.8) becomes,

$$\Delta En = \frac{e^2A^{(0)2}c}{(m^2c^2 + e^2A^{(0)2})^{1/2}} . \qquad (2.16.9)$$

In the limit $eA^{(0)} \gg mc$, Eq. (2.16.9) becomes $\Delta En \rightarrow eA^{(0)}c = \hbar\omega$; using the free photon minimal prescription [1—3] $\hbar\kappa = eA^{(0)}$. In this limit, the photon $\hbar\omega$ is transferred to the electron, and annihilated. This is the high field limit [3]. In the opposite low field limit, $eA^{(0)} \ll mc$, the inverse Faraday effect is,

$$\Delta En \rightarrow \frac{e^2}{m} A^{(0)2} = \frac{(\hbar\omega)^2}{mc^2} = \left(\frac{\hbar\omega}{mc^2}\right)\hbar\omega. \tag{2.16.10}$$

This limit is attained experimentally using visible frequencies, the opposite high field limit using radio frequencies [1—3]. In Eq. (2.16.10), the inverse Faraday effect is seen to be the square of the quantum of electromagnetic energy (i.e., photon squared) divided by the electronic rest energy, mc^2; and is simply the energy transferred inelastically ($\hbar\omega/mc^2 < 1$) in photon-electron collisions. The existence of the effect was first inferred thermodynamically and phenomenologically by Pershan [32], and it was first demonstrated empirically using the induction due to $\boldsymbol{B}^{(3)}$ [13]. If $B^{(0)} = \kappa A^{(0)}$ and if $\boldsymbol{B}^{(3)*} = B^{(0)}\boldsymbol{e}^{(3)*} = \nabla \times \boldsymbol{A}$ in Eq. (2.16.1), then $A^{(0)}$ is the magnitude of $\boldsymbol{A}$; and $B^{(0)} = eA^{(0)2}/\hbar$. Therefore,

$$B^{(0)} = |\boldsymbol{B}^{(3)}| = |\nabla \times \boldsymbol{A}| = \frac{e}{\hbar} A^{(0)2} = \kappa A^{(0)} = \frac{\omega}{c} A^{(0)}, \tag{2.16.11}$$

and in regions where $\boldsymbol{B}^{(3)}$ and $\boldsymbol{A}$ are non-zero they are related by

$$B^{(0)} = |\nabla \times \boldsymbol{A}| = |\nabla \times (\boldsymbol{A} + \nabla\phi)|. \tag{2.16.12}$$

The flux density $B^{(0)}$ creates in addition an inverse Faraday effect in regions where $\boldsymbol{B}^{(3)*}$ and $\boldsymbol{A}$ are zero. This effect is due to $\boldsymbol{B}^{(3)*} = -i\,(e/\hbar)\,(\nabla\phi) \times (\nabla\phi)^*$. The total flux density present in both regions is however, still $B^{(0)}$. In order to see this effect experimentally the standard inverse Faraday experiment [13—21] is modified by excluding the electromagnetic field from direct contact with the electrons. For example, a laser beam in an optical fibre is directed through an electron beam and the inverse induction measured with an induction coil. Observation of this effect would prove action at a distance in electromagnetism, and by implication, gravitation [33].

The total magnetic flux density present is always given by the curl of a sum of potential functions,

$$\boldsymbol{B}^{(3)} = \nabla \times (\boldsymbol{A} + \boldsymbol{A}_1 + \ldots) = \nabla \times \boldsymbol{A} . \tag{2.16.13}$$

When $\boldsymbol{A}$ and $\boldsymbol{B}^{(3)}$ are both zero, the balance of terms in Eq. (2.16.13) is represented by

$$\boldsymbol{0} = \boldsymbol{0} + \nabla \times \boldsymbol{A}_1 + \ldots \tag{2.16.14}$$

where,

$$\boldsymbol{A}_1 \neq \boldsymbol{0}, \qquad \boldsymbol{A}_1 \times \boldsymbol{A}_1^* \neq \boldsymbol{0} . \tag{2.16.14a}$$

Therefore $\nabla \times \boldsymbol{A}_1$ is always zero, but $-\boldsymbol{A}_1^* \times \boldsymbol{A}_1$ is non-zero; and ∇ is not equal to $ie\boldsymbol{A}_1^*/\hbar$. The net result of the gauge transform is therefore,

$$\boldsymbol{B}^{(3)} \rightarrow \boldsymbol{B}_1^{(3)} \rightarrow \boldsymbol{B}_2^{(3)} \text{ etc.,} \tag{2.16.15}$$

i.e., to topologically transfer $\boldsymbol{B}^{(3)}$ from one region of space-time to another. In so doing, the total magnitude $\boldsymbol{B}^{(3)}$ is conserved by Noether's theorem. The magnitude of the expected inverse Faraday effect is therefore the same, outside or inside the fibre.

Acknowledgments

York University, Toronto and the Indian Statistical Institute, Calcutta, are thanked for visiting professorships, and e mail discussions acknowledged with several colleagues worldwide.

References

[1] M. W. Evans and J.-P. Vigier, *The Enigmatic Photon, Vol. 1: The Field* $\boldsymbol{B}^{(3)}$ (Kluwer Academic, Dordrecht, 1994).

[2] M. W. Evans and J.-P. Vigier, *The Enigmatic Photon, Vol. 2: Non-Abelian Electrodynamics* (Kluwer Academic, Dordrecht, 1995).

[3] M. W. Evans, J.-P. Vigier, S. Roy, and S. Jeffers, *The Enigmatic Photon, Vol. 3: Theory and Practice of the* $\boldsymbol{B}^{(3)}$ *Field* (Kluwer, Dordrecht, 1996).

[4] M. W. Evans, *Physica B* **182**, 227, 237 (1992); **183**, 103 (1993); **190**, 310 (1993); *Physica A* **214**, 605 (1995).

[5] M. W. Evans, *The Photon's Magnetic Field* (World Scientific, Singapore, 1992).

[6] M. W. Evans and S. Kielich, eds., *Modern Nonlinear Optics,* Vols. 85(1), 85(2), 85(3) of *Advances in Chemical Physics,* I. Prigogine and S. A. Rice, eds., (Wiley Interscience, New York, 1993).

[7] M. W. Evans, *Found. Phys. Lett.* **7**, 76, 209, 379, 467, 577 (1994); **8**, 63, 83, 187, 363, 385 (1995); *Found. Phys.* **24**, 892, 1519, 1671 (1994); **25**, 175, 383 (1995).

[8] A. A. Hasanein and M. W. Evans, *The Photomagneton in Quantum Field Theory* (World Scientific, Singapore, 1994).

[9] M. W. Evans and S. Roy, *Found. Phys.*, submitted for publication; M. W. Evans, S. Roy and S. Jeffers, *Il Nuovo Cim., D* in press.

[10] M. W. Evans, *Found. Phys. Lett.*, in press and submitted.

[11] M. W. Evans, Found. Phys. and *Apeiron*, submitted for publication.

[12] M. W. Evans, J.-P. Vigier and S. Roy, eds. *The Enigmatic Photon, Vol. 4, New Directions* (Kluwer Academic, Dordrecht, 1998).

[13] J.-P. van der Ziel, P. S. Pershan, and L. D. Malmstrom, *Phys. Rev. Lett.* **15**, 190 (1965); *Phys. Rev.* **143**, 574 (1966).

[14] J. Deschamps, M. Fitaire, and M. Lagoutte, *Phys. Rev. Lett.* **25**, 1330 (1970); *Rev. Appl. Phys.* **7**, 155 (1972).

[15] W. Happer, *Rev. Mod. Phys.* **44**, 169 (1972).

[16] R. Zawodny, in Ref. 6, Vol 85(1), a review of magneto-optical effects.

[17] G. H. Wagnière, *Linear and Nonlinear Optical Properties of Molecules* (VCH, Basel, 1993).
[18] S. Woźniak, M. W. Evans, and G. Wagnière, *Mol. Phys.* **75**, 81, 99 (1992).
[19] P. W. Atkins, *Molecular Quantum Mechanics*, 2nd edn. (Oxford University Press, Oxford, 1983), reviews the inverse Faraday effect.
[20] P. W. Atkins and M. H. Miller, *Mol. Phys.* **75**, 491, 503 (1968).
[21] T. W. Barrett, H. Wohltjen and A. Snow, *Nature* **301**, 694 (1983).
[22] W. E. Ehrenburg and R. E. Siday, Proc. Phys. Soc., **B62**, 8 (1948); Y. Aharonov and D. Bohm, *Phys. Rev.* **115**, 485 (1959).
[23] M. Peshkin and A. Tonomura, *The Aharonov-Bohm Effect*, Vol. 340 (Lecture Notes in Physics, Springer, Berlin, 1989).
[24] F. Hasselbach and M. Nicklaus, *Phys. Rev.* **48**, 143 (1993).
[25] A. van der Merwe and A. Garuccio, eds., *Waves and Particles in Light and Matter* (Plenum, New York, 1994).
[26] D. Bohm, *Phys. Rev.* **85**, 166, 180 (1952).
[27] D. Bohm, B. J. Hiley and P. N. Kaloreyou, *Phys. Rep.* **144**, 321 (1987).
[28] V. Dvoeglazov, e mail communications.
[29] A. Chubykalo and R. Smirnov-Rueda, *Phys. Rev. E*, in press.
[30] H. Muñera and O. Guzmán, *Found. Phys. Lett.*, submitted for publication.
[31] B. Talin, V. P. Kaftandjan, and L. Klein, *Phys. Rev. A* **11**, 648 (1975).
[32] P. S. Pershan, *Phys. Rev.* **130**, 919 (1963).
[33] M. W. Evans, *Found. Phys. Lett.*, in press.

Paper 17

Maxwell's Vacuum Field — a Rotating Charge

The rotating electric field of a propagating Maxwellian plane wave in vacuo is shown to be a rotating dipole, with one tip fixed at the origin. At the other there is located a rotating charge, whose helical trajectory forms the Evans-Vigier field $\left(\boldsymbol{B}^{(3)}\right)$ in perfect analogy with a solenoid. The length of the dipole is $r_0 = \left(V/(4\pi\alpha)\right)^{1/3}$, where V is the radiation volume and α the fine structure constant. The quantum of electromagnetic radiation (the photon) is, consequently, $\hbar\omega = (4\pi\alpha V)^{3/2}\, e^2/\epsilon_0$, where ϵ_0 is the vacuum permittivity.

Key words: Maxwellian vacuum field, rotating charge, Evans-Vigier field.

17.1 Introduction

The recent emergence [1—12] of the classical Evans-Vigier field, $\boldsymbol{B}^{(3)}$, the spin field of vacuum electromagnetism, has led to the charge quantization condition,

$$\hbar\kappa = eA^{(0)}, \tag{2.17.1}$$

where the de Broglie photon momentum, $\hbar\kappa$, is expressed as $eA^{(0)}$, where e is the charge on the electron and $A^{(0)}$ the scalar potential in vacuo. Here $\hbar$ is the Dirac constant as usual and κ the magnitude of the classical wavevector in vacuo. In this Letter it is shown that Eq. (2.17.1) is a straightforward consequence of the existence of the rotating Maxwellian electric field in the vacuum, and that Eq. (2.17.1), derived independently [6,12], using generalized gauge theory, is a consistent outcome of standard field theory. This proves beyond reasonable doubt that $\boldsymbol{B}^{(3)}$ is also consistent in field theory, and is a novel fundamental property of electromagnetic radiation in the vacuum.

In Sec.17.2, it is shown that Eq. (2.17.1) emerges directly from the link between an electric field strength (V m^{-1}) and a dipole moment. In *S.I.* units, this link is

$$\boldsymbol{\mu} = \epsilon_0 V\boldsymbol{E}, \tag{2.17.2}$$

where V has the units of volume, and where ϵ_0 is the vacuum permittivity. In the vacuum, there is no material polarization, and V is the volume occupied by classical electromagnetic radiation in vacuo. In Sec. 17.3, it is shown that the length of the dipole is

$$r_0 = \frac{1}{\kappa} = \left(\frac{V}{4\pi\alpha}\right)^{1/3} = \frac{\lambda}{2\pi}, \tag{2.17.3}$$

where α is the fine structure constant [13] and λ is the wave length; the quantum of electromagnetic radiation (the photon) is

$$\hbar\omega = \frac{(4\pi\alpha V)^{3/2}e^2}{\epsilon_0}, \tag{2.17.4}$$

where e is the charge on the electron. The photon is therefore proportional to the square of e, and this is the origin of the textbook assertion [14] that the photon is uncharged. This assertion is seen through Eq. (2.17.4) to be true only in the narrowest of senses: the photon only *appears* to be uncharged because it is proportional to the *square* of the electronic charge. Section 17.4 is a discussion of these findings.

17.2 Derivation of the Charge Quantization Condition

An electric dipole moment is a separation of opposite charge [15]. An electric field is generated by separated opposite charge, or by a single charge, as in Coulomb's law. The elementary charge is that on the electron, of magnitude e, and so

$$\boldsymbol{E}^{(1)} = \frac{\boldsymbol{\mu}^{(1)}}{\left(\epsilon_0 V\right)} = \frac{e\boldsymbol{r}^{(1)}}{\epsilon_0 V}, \tag{2.17.5}$$

where $\boldsymbol{E}^{(1)}$ is Maxwell's rotating electric field in electromagnetic radiation propagating through the vacuum [15]. In Eq. (2.17.5), $r^{(0)} := |\boldsymbol{r}^{(1)}|$ is the length of the dipole formed by these separated charges in vacuo. If V is the radiation volume, it is easily checked that Eq. (2.17.5) is dimensionally and physically consistent in the theory of electrodynamics [15]. The electromagnetic field propagates in the axis perpendicular to the plane , and so the negative rotating charge draws out a helical path through the vacuum. In perfect analogy with the solenoid, this movement produces a magnetic flux density $\boldsymbol{B}^{(3)}$ in the axis of propagation, and this is the Evans-Vigier field [6]. The origin of $\boldsymbol{B}^{(3)}$ becomes perfectly clear in classical electrodynamics.

It has been shown elsewhere [6,12] that the existence of $\boldsymbol{B}^{(3)}$ means that

$$\hbar\kappa = eA^{(0)}, \tag{2.17.6}$$

where $\hbar\kappa$ is the magnitude of the de Broglie photon momentum. Equation (2.17.6) appears at first to be unorthodox, in that it expresses $\hbar\kappa$ in terms of e, the charge on the electron. However, Eq. (2.17.6) is easily derived from Eq. (2.17.5) as follows.

From Eq. (2.17.5), the scalar magnitude of the rotating Maxwellian $\boldsymbol{E}^{(1)}$ (V m^{-1}) is

$$E^{(0)} = \frac{er^{(0)}}{\epsilon_0 V} . \tag{2.17.7}$$

The classical electromagnetic energy in the volume V is [14]

$$En = \epsilon_0 E^{(0)2} V, \tag{2.17.8}$$

where

$$V := \int_0^V dV, \tag{2.17.9}$$

and so the energy in volume V is expressible in terms of the radius $r^{(0)}$,

$$En = eE^{(0)} r^{(0)} = \frac{e^2 r^{(0)2}}{\epsilon_0 V} . \tag{2.17.10}$$

Thus far, the development has been entirely classical. Using the classical relation between $A^{(0)}$ and $E^{(0)}$ [6],

$$A^{(0)} = \frac{E^{(0)}}{\omega} , \tag{2.17.11}$$

it is found that

$$En = \left(eA^{(0)}r^{(0)}\right)\omega, \tag{2.17.12}$$

i.e., electromagnetic energy is proportional to electromagnetic angular frequency as radiation propagates in vacuo. Provided we make the identity

$$\hbar := eA^{(0)}r^{(0)}, \tag{2.17.13}$$

Eq. (2.17.12) is the Planck-Einstein relation of quantum theory, the light quantum hypothesis. *It has been shown that the Light Quantum hypothesis has a purely classical origin.*

Finally, the identification of $r^{(0)}$ as κ^{-1} results in the charge quantization condition that we are seeking to derive. The length of the rotating dipole, and therefore of the rotating Maxwellian electric field, is therefore the inverse of the wavevector magnitude of the radiation. *We have identified the origin of the Planck constant itself.*

17.3 The Dipole Radius and Photon

Using Eq. (2.17.6) with the equation

$$A^{(0)} = \left(\frac{c}{\epsilon_0 \omega^2 V}\right) e, \tag{2.17.14}$$

the quantum of electromagnetic energy, $\hbar\omega$, is defined as

$$\hbar\omega = \frac{e^2}{\epsilon_0 \kappa^2 V}, \tag{2.17.15}$$

and is proportional to e^2. This is why the particulate photon is *uncharged.* This result can now be expressed in terms of the fine structure constant [13, 16],

$$\alpha := \frac{e^2}{4\pi c \epsilon_0 \hbar} . \tag{2.17.16}$$

From Eqs. (2.17.15) and (2.17.16), the fine structure constant becomes expressible as

$$\alpha = \frac{V}{4\pi r_0^3} , \tag{2.17.17}$$

and so the radius of the rotating dipole is

$$r_0 = \frac{1}{\kappa} = \left(\frac{V}{4\pi\alpha} \right)^{1/3} = \frac{\lambda}{2\pi}, \tag{2.17.18}$$

so that the photon defined in equation (2.17.15) becomes expressible as

$$\hbar\omega = \left(4\pi\alpha V \right)^{3/2} \frac{e^2}{\epsilon_0} . \tag{2.17.19}$$

17.4 Discussion

The rotating dipole is consistent with special relativity and with Maxwell's equations, because it is derived simply by re-expressing (Eq. (2.17.2)) the usual rotating electric field in terms of separated charge. The positive charge is located at the origin because the rotating electric field rotates around the origin, as described for example by Jackson [15]. If the rotating electric charge is identified with the charge on the electron, e, and the dipole radius is denoted $r^{(0)}$, Eq. (2.17.12) follows from *classical* electrodynamics. However, Eq. (2.17.12) is the basis of *quantum* theory, and is the light quantum hypothesis suggested originally by Planck in Nov., 1900. Equation (2.17.12) shows that in classical electrodynamics, the energy En of a plane wave in vacuo is proportional to its angular frequency, ω, through $eA^{(0)}r^{(0)}$. The Planck-Einstein relation of the quantum theory requires this to be $\hbar$, and so

$$\hbar = eA^{(0)}r^{(0)} = \frac{eA^{(0)}}{\kappa}, \qquad (2.17.20)$$

which is the charge quantization condition [6] provided that the radius $r^{(0)}$ is the inverse of the magnitude of the wavevector. The quantity $eA^{(0)}/\kappa$ is therefore a constant angular momentum in *classical* electrodynamics for the propagating plane wave in vacuo, and is identified as the angular momentum, $\hbar$, of one photon, the latter being the quantum of electromagnetic radiation. The quantum theory is therefore classical in origin, and we have shown that Planck's *assertion* that En must be proportional to ω is, in fact, a logical outcome of classical electrodynamics. The photon is a classical entity.

Furthermore, our innocent expression of $\boldsymbol{E}^{(1)}$ as an electric dipole moment results immediately in the Evans-Vigier field $\boldsymbol{B}^{(3)}$ [1—12], which is shown to be an outcome of helical charge motion as in a solenoid. In this respect the positive charge is fixed at the origin, and does not cancel out the induction of $\boldsymbol{B}^{(3)}$ because the positive charge is not rotating, and therefore does not induce a $\boldsymbol{B}^{(3)}$ in the opposite direction.

The usual idea of a *photon* as being uncharged [14], and therefore its own anti-particle, is shown by Eq. (2.17.19) to have the narrowest of meanings. The *photon* is a classical amount of energy, which is proportional to the *square* of the charge on the electron, and is not *uncharged.* The charge quantization condition shows that the origin of the Dirac constant $\hbar$, (or of the Planck constant h) is the electronic charge e multiplied by $A^{(0)}$, which is, within a constant c, the scalar potential [15] of the classical electromagnetic wave. Re-expressing $A^{(0)}$ through Eq. (2.17.15) leads to Eq. (2.17.17), which expresses the fine structure constant of quantum electrodynamics as a simple ratio of volumes. The fine structure constant is the ratio of the volume, V, used in Eq. (2.17.2) to the volume of the sphere defined by the radius $r^{(0)}$, the length of the dipole. This spherical volume is

$$V_0 = \frac{4}{3}\pi r^{(0)3} , \tag{2.17.21}$$

and so the fine structure constant is the ratio

$$\alpha = \frac{1}{3}\frac{V}{V_0} . \tag{2.17.22}$$

Equation (2.17.18) shows that the magnitude of the classical Maxwellian wavevector is defined by the volume V through

$$\kappa = \left(\frac{4\pi\alpha}{V}\right)^{1/3} , \tag{2.17.23}$$

and because α is a universal constant [13], the de Broglie photon momentum becomes

$$p = \hbar\kappa = \left(\frac{4\pi\alpha\hbar^3}{V}\right)^{1/3} = \left(\frac{e^2\hbar^2}{\epsilon_0 c V}\right)^{1/3} , \tag{2.17.24}$$

and the Planck-Einstein photon becomes expressible as

$$En = \hbar\omega = \hbar\kappa c = \left(\frac{e^2\hbar^2 c^2}{\epsilon_0 V}\right)^{1/3} . \tag{2.17.25}$$

These equations show that the quantum of energy, $\hbar\omega$, and the quantum of radiation momentum, $\hbar\kappa$, are both defined in terms only of V and the universal fine structure constant α. This is consistent with the fact that they are fundamental quanta of energy and momentum, and with the fact that the magnitudes of these quanta vary *only* with the volume V introduced in Eq. (2.17.2). On the most fundamental level in this Maxwellian theory, everything depends ultimately on e, the charge on the electron, and the constant h is a consequence of the existence of e. In generalized gauge theory [1—12], from which the charge quantization condition emerges [10—12], e simultaneously plays the role of a gauge scaling factor, i.e., is geometrical in nature, so that an angular momentum such as h becomes understandable in terms of a geometrical entity, e. We know the latter much more familiarly as the elementary unit of charge, an elementary geometrical measure of the known universe.

Finally, the concept of photon mass as envisaged [17] by de Broglie, Bohm, Vigier, and co-workers appears to be consistent with the rotating Maxwellian dipole because if there is mass involved in this motion, *it must be concentrated at the origin*, because the center of mass of the rotating dipole must be there because it is rotating about that point. If the mass were distributed, for example if the negative charge had some mass as well as the positive charge, the center of mass would be somewhere between the two charges, and the dipole could not rotate about one point. The particulate photon can therefore be envisaged with a mass concentrated at the origin. This is still a *classical* picture, because h has been identified as being classical in nature. This picture becomes consistent with special relativity if we use the de Broglie guidance theorem,

$$m_r = \frac{\hbar\omega_r}{c^2} = \frac{\hbar\kappa_r}{c} = \left(\frac{e^2\hbar^2}{\epsilon_0 c^4 V_r}\right)^{1/3}, \tag{2.17.26}$$

and so express the photon rest mass, m_r, in terms of the rest volume V_r. The introduction of mass is necessary because the Evans-Vigier field, $\boldsymbol{B}^{(3)}$, is *longitudinal*, so that the Wigner little group for a massless particle, E(2), becomes untenable [6]. The existence of $\boldsymbol{B}^{(3)}$ shows immediately that the particulate photon is concomitant with *three* degrees of space polarization, which, in the circular basis [6], are (1), (2) and (3). If the theory of special relativity is accepted therefore, the photon cannot be massless because of the existence of $\boldsymbol{B}^{(3)}$.

Acknowledgments

It is a pleasure to acknowledge discussions with several colleagues among whom are: Gareth J. Evans, Stanley Jeffers, Malcolm Mac Gregor, Mikhail Novikov, Sisir Roy, and Jean-Pierre Vigier.

References

[1] M. W. Evans, *Physica B* **182**, 227, 237 (1992); **183**, 103 (1993).
[2] M. W. Evans, *Physica B* **190**, 310 (1993).
[3] M. W. Evans in A. van der Merwe and A. Garuccio eds., *Waves and Particles in Light and Matter* (Plenum, New York, 1994), The de Broglie Centennial Volume.
[4] M. W. Evans, *The Photon's Magnetic Field* (World Scientific, Singapore, 1992).
[5] in M. W. Evans, and S. Kielich, eds., *Modern Nonlinear Optics*, Vol. 85(2) of *Advances in Chemical Physics*, I. Prigogine and S. A. Rice, eds., (Wiley Interscience, New York, 1993).
[6] M. W. Evans and J.-P. Vigier, *The Enigmatic Photon, Volume 1, The Field* $\boldsymbol{B}^{(3)}$ (Kluwer, Dordrecht, 1994); *The Enigmatic Photon, Volume 2, Non-Abelian Electrodynamics* (in press.); M. W. Evans, J.-P. Vigier with S. Jeffers and S. Roy, *The Enigmatic Photon, Volume 3, Collected Papers on* $\boldsymbol{B}^{(3)}$ (Kluwer Academic, Dordrecht, 1996).
[7] A. A. Hasanein and M. W. Evans, *The Photomagneton and Quantum Field Theory*, Volume one of *Quantum Chemistry.* (World Scientific, Singapore, 1994).

[8] M. W. Evans, *Mod. Phys. Lett.* **7**, 1247 (1993).
[9] M. W. Evans, *Found. Phys. Lett.* **7**, 67 (1994).
[10] M. W. Evans, ibid., in press, 1994.
[11] M. W. Evans, *Found. Phys.* in press, 1994.
[12] M. W. Evans, *Found. Phys. Lett.* in press, 1994/1995.
[13] P. W. Atkins, *Molecular Quantum Mechanics* 2nd. edn. (Oxford University Press, Oxford, 1983).
[14] L. H. Ryder, *Elementary Particles and Symmetries* (Gordon and Breach, London, 1986).
[15] J. D. Jackson, *Classical Dynamics.* (Wiley, New York, 1962).
[16] P. Cornille, in Ref. 3; L. H. Ryder, *Quantum Field Theory*, 2nd. edn. (Cambridge University Press, Cambridge, 1987).
[17] The de Broglie Centennial Volume, Ref. 3, contains several excellent accounts of the de Broglie-Bohm-Vigier (Paris School) Theory of Quantum Mechanics.

Paper 18

Dipole Model for the Photon and the Evans-Vigier Field

The fundamental magnetizing field of light, $\boldsymbol{B}^{(3)}$, is expressed in terms of vector cross products of oscillating electric and magnetic dipole moments of vacuum electromagnetic radiation. The consequences are developed with reference to a new dipole model of the photon recently suggested by Mac Gregor [16]. The similarities and differences in the two theories are analyzed briefly.

Key words: $\boldsymbol{B}^{(3)}$ Field, Dipole Model.

18.1 Introduction

The photon as particle is usually considered to be without mass: light, after all, is assumed to travel at the speed of light, which is c in the vacuum. The recent emergence [1—12] of the Evans-Vigier field $\boldsymbol{B}^{(3)}$ has shown, however, that this idea cannot be physically meaningful, essentially because $\boldsymbol{B}^{(3)}$ indicates a third degree of polarization which is forbidden for a massless particle [6]. The unphysical nature of a massless particle was first inferred by Wigner [13] who showed that the little group in such a case is *E(2)*. This is the Euclidean group of rotations and translations simultaneously taking place *in a plane*. Such a group has no physical meaning [14]. Therefore the idea of a massless particle is physically obscure in classical special relativity, but is paradoxically accepted by the contemporary majority. Since $\boldsymbol{B}^{(3)}$ is observable directly in magneto-optical phenomena, such as the inverse Faraday effect [15], Wigner's deduction receives experimental verification through $\boldsymbol{B}^{(3)}$. The existence of this field in the vacuum has now been inferred from the Hamiltonian principle of least action and from the Dirac equation [6], and it is therefore a new, fundamental, property of light. In consequence, the photon, as particle, is concomitant with physically meaningful field components in three polarizations, denoted by $\boldsymbol{B}^{(1)}$, $\boldsymbol{B}^{(2)}$ and $\boldsymbol{B}^{(3)}$ in a complex circular representation, (1), (2) and (3), of three dimensional space [6]. Two of these axes are transverse (perpendicular) to the direction of propagation and the third, (3), is longitudinal, meaning that $\boldsymbol{B}^{(3)}$ is an axial vector directed in the (3) axis.

The traditional view recognizes the existence of the conjugate product $\boldsymbol{B}^{(1)} \times \boldsymbol{B}^{(2)}$ in, for example, the inverse Faraday effect [15], but the all important extra inference,

$$\boldsymbol{B}^{(1)} \times \boldsymbol{B}^{(2)} = iB^{(0)}\boldsymbol{B}^{(3)*}, \qquad \boldsymbol{B}^{(2)} \times \boldsymbol{B}^{(3)} = iB^{(0)}\boldsymbol{B}^{(1)*},$$
$$\boldsymbol{B}^{(3)} \times \boldsymbol{B}^{(1)} = iB^{(0)}\boldsymbol{B}^{(2)*}, \tag{2.18.1}$$

was achieved only recently [1]. Here $\boldsymbol{B}^{(1)} = \boldsymbol{B}^{(2)*}$ is the complex magnetic plane wave in vacuo, $B^{(0)}$ the scalar amplitude of the magnetic flux density carried by the wave, and where $\boldsymbol{B}^{(3)} = \boldsymbol{B}^{(3)*}$ is a real, phase free, spin field. In the traditional view, $\boldsymbol{B}^{(3)}$ is unconsidered, the vacuum Maxwell equations are applied only to plane *waves*. Due to the Lie algebra [1], however, essentially the algebra of the *O(3)* rotation group, the very existence of the plane waves $\boldsymbol{B}^{(1)}$ and $\boldsymbol{B}^{(2)}$ means the existence of $\boldsymbol{B}^{(3)}$, which is real and physical, and which magnetizes matter in the inverse Faraday effect [15]. The conjugate product $\boldsymbol{B}^{(1)} \times \boldsymbol{B}^{(2)}$ can also be interpreted in a meaningful, physical, way because $\boldsymbol{T}_V = -\boldsymbol{B}^{(1)} \times \boldsymbol{B}^{(2)}/\mu_0$ is the torque per unit volume carried by electromagnetic radiation in vacuo. This is pure imaginary, i.e., has no real part because the real angular momentum density of the radiation is constant (the angular momentum of one photon is $\hbar$) and therefore its real time derivative is zero. The quantity $\boldsymbol{T}_V/c$ is the antisymmetric part of the tensor of light intensity [4] a tensor whose scalar part is, in this notation,

$$I_0 = \frac{1}{c}|\boldsymbol{T}_V| = \frac{B^{(0)2}}{\mu_0 c} . \tag{2.18.2}$$

Therefore $\boldsymbol{B}^{(3)}$ and $\boldsymbol{T}_V$ can be expressed in terms of the S_3 Stokes parameter [3].

Although $\boldsymbol{T}_V$ is pure imaginary in the vacuum, it causes magnetization in the inverse Faraday effect [15] through the imaginary part of material hyperpolarizability, in the simplest case, one electron hyperpolarizability [6]. Therefore both $\boldsymbol{B}^{(3)}$ and $\boldsymbol{T}_V$ play a role in the correct, relativistic, treatment of the inverse Faraday effect, and this can be demonstrated from the principle of least action by using the relativistic Hamilton-Jacobi equation of one electron in the circularly polarized electromagnetic field. Therefore $\boldsymbol{B}^{(3)}$ is inferred both experimentally (through the observation [15] of the inverse Faraday effect) and theoretically (from the relativistic principle of least action).

In this Letter, the vacuum $\boldsymbol{B}^{(3)}$ field is expressed through conjugate products of the complex magnetic and electric dipole moments of the radiation in vacuo (Sec. 18.2). Section 18.3 compares this result with the theory of Mac Gregor [16], who has developed an interesting dipole formulation of the photon, both in terms of charge and mass.

18.2 Double Dipole Expressions for $B^{(3)}$

The field $\boldsymbol{B}^{(3)}$ emerges from the free space electromagnetic torque density as follows

$$\boldsymbol{B}^{(3)} = \frac{i\mu_0}{B^{(0)}}\boldsymbol{T}_V, \tag{2.18.3}$$

and although $\boldsymbol{T}_V$ has no real part, $\boldsymbol{B}^{(3)}$ is real and physical. Recently, Mac Gregor has developed a model of the photon based on the concept of symmetric particle-antiparticle excitation [16] of the vacuum state. The electrodynamic part of this model reduces to a picture of the photon as a rotating dipole with zero net charge. The Mac Gregor electric dipole rotates in the plane orthogonal to the axis of propagation, $\boldsymbol{e}^{(3)}$. Antecedents of this picture were traced [16] to Bateman [17], Bonnor [18], and J. J. Thomson [19]. Bonnor [18] added masses to the charges, and developed the theory in the context of special and general relativity, showing that the electromagnetic energy density of an electric dipole traveling at the speed of light is finite.

In this section, we develop an analogous theory for $\boldsymbol{B}^{(3)}$ in terms of magnetic and electric dipole moments of the vacuum radiation. The cyclic relations,

$$\boldsymbol{m}^{(1)} \times \boldsymbol{m}^{(2)} = i\mu_0 B^{(0)}\boldsymbol{m}^{(3)*}, \tag{2.18.4}$$

emerge directly from the *S.I.* relation,

$$\boldsymbol{m} = \frac{\boldsymbol{B}}{\mu_0}, \tag{2.18.5}$$

where $\boldsymbol{m}$ is a magnetic dipole moment. Therefore $\boldsymbol{m}^{(1)}$, $\boldsymbol{m}^{(2)}$ and $\boldsymbol{m}^{(3)}$ are magnetic dipole moments of the radiation *itself*, and should not be confused with material dipole moments in matter. The reason is that Eq. (2.18.4) is one for vacuum propagation of light.

A similar relation for electric dipole moments of the radiation can be deduced from the *S.I.* relation,

$$\boldsymbol{\mu} = \epsilon_0 V \boldsymbol{E} , \tag{2.18.6}$$

where V is the radiation volume and ϵ_0 the vacuum permittivity. Equation (2.18.6) follows directly from the relation between polarization and electric dipole moment density,

$$\boldsymbol{P} = \frac{\boldsymbol{\mu}}{V} . \tag{2.18.7}$$

If $\boldsymbol{E}^{(1)} = \boldsymbol{E}^{(2)*}$ is the electric field strength (V m^{-1}) of the vacuum plane wave, and $E^{(0)}$ its amplitude, we have [6],

$$\boldsymbol{B}^{(3)*} = \frac{1}{icE^{(0)}} \boldsymbol{E}^{(1)} \times \boldsymbol{E}^{(2)}, \tag{2.18.8}$$

and so

$$\boldsymbol{B}^{(3)*} = -i\frac{\epsilon_0}{\mu_0} \cdot \frac{1}{B^{(0)}} \frac{\boldsymbol{\mu}^{(1)}}{V} \times \frac{\boldsymbol{\mu}^{(2)}}{V} , \tag{2.18.9}$$

which expresses the Evans-Vigier field in terms of the cross product of electric dipole moment densities of the vacuum radiation itself.

18.3 Discussion

Equation (2.18.9) is an expression of the $\boldsymbol{B}^{(3)}$ field in terms of,

$$\boldsymbol{\mu}^{(1)} = \epsilon_0 V \frac{E^{(0)}}{\sqrt{2}} (\boldsymbol{i} - i\boldsymbol{j}) e^{i\phi}, \tag{2.18.10}$$

which is itself a traveling plane wave in the vacuum. An electric dipole moment can be analyzed in terms of positive and negative charges separated by a distance $\boldsymbol{r}$, and because the vector is rotating as the light beam propagates in the $\boldsymbol{e}^{(3)}$ axis, the $\boldsymbol{B}^{(3)}$ field is formed from this motion. The rotations of the positive and negative charges reinforce each other in the creation of the vacuum $\boldsymbol{B}^{(3)}$ by Eq. (2.18.9), and $\boldsymbol{B}^{(3)}$ itself is relativistically invariant. This classical analysis, which is a direct result of the vacuum Maxwell equations, is similar to that proposed by Mac Gregor [16], who points out that the classical electromagnetic field is always the result of the movement of charge by Ampère's hypothesis [20]. The simple analysis of Sec. (18.2) shows that the field can be expressed in terms of either magnetic or electric dipole moments, and therefore in terms of separated charges. The field amplitudes $E^{(0)}$ and $B^{(0)}$ are negative under charge conjugation, $\hat{C}$, [21], and in the quantized field the photon is also negative under $\hat{C}$,

$$\hat{C}(\gamma) = -\gamma . \tag{2.18.11}$$

Therefore the particulate photon is always concomitant in vacuo with $\hat{C}$ negative electric and magnetic fields and dipoles.

Equation (2.18.1) defines a novel magnetic field, $\boldsymbol{B}^{(3)}$ [6], the Evans-Vigier field, that has a physical existence in the vacuum orthogonal to the plane defined by $\boldsymbol{B}^{(1)}$ or $\boldsymbol{E}^{(1)}$; or by $\boldsymbol{m}^{(1)}$ for $\boldsymbol{\mu}^{(1)}$ in the circular basis [1—12],

$$\boldsymbol{e}^{(1)} \times \boldsymbol{e}^{(2)} = i\boldsymbol{e}^{(3)*}, \text{ et cyclicum,} \tag{2.18.12}$$

for three dimensional space [6]. The gauge group [22] of vacuum electrodynamics must in consequence [6,16] become *O(3)*, and fundamental gauge theory [6,22] leads directly to,

$$\boldsymbol{B}^{(3)*} = -i\frac{e}{\hbar}\boldsymbol{A}^{(1)} \times \boldsymbol{A}^{(2)}, \tag{2.18.13}$$

where $\boldsymbol{A}^{(1)} = \boldsymbol{A}^{(2)*}$ is the vector potential defined by,

$$\boldsymbol{B}^{(1)} = \nabla \times \boldsymbol{A}^{(1)}. \tag{2.18.14}$$

Equation (2.18.13) leads to the charge quantization condition [6],

$$\hbar\kappa = eA^{(0)}, \tag{2.18.15}$$

where $\hbar\kappa$ is the magnitude of the linear momentum in vacuo of the free photon, and where $A^{(0)}$ is the scalar magnitude of $\boldsymbol{A}^{(1)}$. Equation (2.18.15) [6] shows that the magnitude (e) of the electronic charge is also the scaling factor of the *O(3)* gauge group, and is defined by the ratio $\hbar\kappa/A^{(0)}$. This is a direct consequence of the existence of $\boldsymbol{B}^{(3)}$ through Eqs. such as (2.18.1) or (2.18.9), equations which spring from the three dimensional nature of space itself.

In Mac Gregor's terminology [16], e and $A^{(0)}$ form a neutral pair, and are both $\hat{C}$ negative quantities. Therefore the conventional view [20—22] that the photon is *uncharged* is narrowly defined, both in Mac Gregor's analysis, and in that given here.

The physical interpretation of Eq. (2.18.10) shows the clockwise rotation of $\boldsymbol{\mu}^{(1)}$ for a right circularly polarized electromagnetic field in vacuo, propagating towards the observer [20] with negative helicity. One tip of the rotating dipole describes a circle around the origin if we neglect the forward motion of the wave. At the origin, the other tip of the dipole is fixed. If the moving tip is thought of as negatively charged, that charge moves on a helix as the electromagnetic wave propagates, and this is precisely analogous with the current in a solenoid. It is clear that the Evans-Vigier field is induced in the propagation axis (*Z*) as a result of this motion of $\boldsymbol{\mu}^{(1)}$. If the motion is anticlockwise, $\boldsymbol{B}^{(3)}$ reverses sign and the wave becomes left circularly polarized with positive helicity [20]. This is represented mathematically by,

$$\boldsymbol{\mu}^{(1)} = \frac{\mu^{(0)}}{\sqrt{2}}(\boldsymbol{i} \mp i\boldsymbol{j})e^{i\phi}, \qquad \phi := \omega t - \boldsymbol{\kappa}\cdot\boldsymbol{r}, \tag{2.18.16}$$

$$\mu^{(0)} := \epsilon_0 V E^{(0)},$$

where ω is the angular frequency of rotation at an instant t, and $\boldsymbol{\kappa}$ is the wave vector at position $\boldsymbol{r}$. Equation (2.18.16) represents a rotating and simultaneously translating electric dipole moment.

This is similar to the Mac Gregor theory [16] in so far as the dipole rotates, but Mac Gregor places the positive and negative charges of the dipole symmetrically about the origin. For this reason, the rotation of the dipole cannot produce a $\boldsymbol{B}^{(3)}$ field in Mac Gregor's theory, but can in our theory. The conventional $\boldsymbol{E}^{(1)}$ field of classical vacuum electrodynamics [20] has simply been replaced by $\boldsymbol{\mu}^{(1)}/(\epsilon_0 V)$, which obviously has the same units of V m^{-1} in *S.I.* This innocent replacement allows clarification of the physical origin of the Evans-Vigier field, which becomes recognizable as precisely analogous with the magnetic field produced by a solenoid. It also becomes clear that there is no longitudinal electric field, as deduced elsewhere [1-12], because there is none in a solenoid.

Therefore $\boldsymbol{\mu}^{(1)}$ rotates *with one end fixed*, the other tip traces out a helix as the electromagnetic wave propagates in vacuo. This is the precise physical origin of the Evans-Vigier field, observable experimentally in magneto-optical effects such as the inverse Faraday effect [15]. The interesting concepts of Mac Gregor [16] can therefore be used for $\boldsymbol{B}^{(3)}$ provided that one tip of the rotating dipole is fixed at the origin, which lies on the Z axis of propagation, while the other tip rotates. The mass distribution [16] must be adjusted accordingly so that mass is concentrated at or near the origin, and this is precisely the concept used by de Broglie, Vigier and Bohm [6] in the realist view of quantum mechanics, in which the photon has mass.

Acknowledgments

It is a pleasure to acknowledge many Internet discussions with several colleagues, including: Stanley Jeffers, Mikhail Novikov, Sisir Roy and Mark Silverman. Mail correspondence with Malcolm Mac Gregor and Jean-Pierre Vigier was of key importance to the development of this theory. Some helpful remarks by O. Costa de Beauregard are acknowledged with thanks.

References

[1] M. W. Evans, *Physica B* **182**, 227, 237 (1992).
[2] M. W. Evans, *Physica B* **183**, 103 (1993); **190**, 310 (1993).
[3] M. W. Evans, *The Photon's Magnetic Field.* (World Scientific, Singapore, 1992).
[4] in M. W. Evans, and S. Kielich, eds., *Modern Nonlinear Optics*, Vol. 85(2) of *Advances in Chemical Physics*, I. Prigogine and S. A. Rice, eds., (Wiley Interscience, New York, 1993).
[5] A. A. Hasanein and M. W. Evans, *The Photomagneton and Quantum Field Theory.*, volume one of *Quantum Chemistry* (World Scientific, Singapore, 1994).
[6] M. W. Evans and J.-P. Vigier, *The Enigmatic Photon, Volume 1: The Field* $\boldsymbol{B}^{(3)}$. (Kluwer, Dordrecht, 1994); *The Enigmatic Photon, Volume 2: Non-Abelian Electrodynamics.* (Kluwer Academic, Dordrecht, 1995).
[7] M. W. Evans in A. van der Merwe and A. Garuccio eds., *Waves and Particles in Light and Matter* (Plenum, New York, 1994), the de Broglie Centennial volume.
[8] M. W. Evans, *Mod. Phys. Lett.* **7**, 1247 (1993).
[9] M. W. Evans, *Found. Phys. Lett.* **7**, 67 (1994).
[10] M. W. Evans, *Found. Phys. Lett.* in press, (1994).
[11] M. W. Evans, *Found. Phys.* in press (1994).
[12] M. W. Evans, *Found. Phys.* in press (1994).
[13] E. P. Wigner, *Ann. Math.* **40**, 149 (1939).

[14] S. Weinberg, *Phys. Rev.* **134B**, 882 (1964).

[15] A. Piekara and S. Kielich, Arch. Sci., **11**, 304 (1958); P. S. Pershan, *Phys. Rev.* **130**, 919 (1963); J. P. van der Ziel, P. S. Pershan, and L. D. Malmstrom, *Phys. Rev.* **143**, 574 (1966); J. Deschamps, M. Fitaire, and M. Lagoutte, *Phys. Rev. Lett.* **25**, 1330 (1970); *Rev. Appl. Phys.* **7**, 155 (1972); S. Woźniak, M. W. Evans, and G. Wagnière, *Mol. Phys.* **75**, 81, 99 (1992); reviewed by R. Zawodny in Ref. 4, Vol. 85(1).

[16] M. H. Mac Gregor, *Found. Phys.* in press.

[17] H. Bateman, *Phil. Mag.* **46**, 977 (1923).

[18] W. B. Bonnor, *Int. J. Theor. Phys.* **2**, 373 (1969); **3**, 57 (1970).

[19] M. H. Mac Gregor, *The Enigmatic Electron* (Kluwer, Dordrecht, 1992).

[20] J. D. Jackson, *Classical Electrodynamics* (Wiley, New York, 1962).

[21] L. H. Ryder, *Elementary Particles and Symmetries* (Gordon and Breach, London, 1986).

[22] L. H. Ryder, *Quantum Field Theory*, 2nd edn. (Cambridge University Press, Cambridge, 1987).

Paper 19

Electromagnetism in Curved Space-time

A suggestion is developed for a theory of electromagnetism in curved space-time, a theory based on a novel *anti*symmetric Ricci tensor which is postulated to be directly proportional to the $G_{\mu\nu}$ tensor of Evans and Vigier, and which therefore deals self-consistently with the experimentally observable $\boldsymbol{B}^{(3)}$ field of magneto-optics.

Key words: Electromagnetism; general relativity; $\boldsymbol{B}^{(3)}$ field.

In this note, a brief summary is given of the essentials of a novel theory [1—3] of electromagnetism, a theory necessitated by the tiny experimental magneto-optic effects [4—8] which need for their self-consistent explanation the $\boldsymbol{B}^{(3)}$ field [9—15] in *curved* space-time. The essence of our argument here is that $\boldsymbol{B}^{(3)}$ can be obtained straightforwardly from the Riemann tensor [6—18] by using the contraction indicated by

$$R^{(A)}_{\mu\nu} := R^{\lambda}_{\lambda\mu\nu} \, . \tag{2.19.1}$$

This produces an *antisymmetric* Ricci tensor $R^{(A)}_{\mu\nu}$ to which the electromagnetic field tensor $G_{\mu\nu}$ introduced by Evans and Vigier [9,10] is directly proportional,

$$R_{\mu\nu}^{(A)} = \frac{e}{\hbar} G_{\mu\nu} \ . \tag{2.19.2}$$

Here $e/\hbar$ is a universal constant, the ratio of the quanta of charge and action (or angular momentum). The Ricci tensor $R_{\mu\nu}^{(A)}$ is defined through affine parameters (Christoffel symbols) in the usual way [16—18],

$$R_{\mu\nu}^{(A)} = \partial_\mu \Gamma^\lambda_{\lambda\nu} - \partial_\nu \Gamma^\lambda_{\lambda\mu} + \Gamma^\rho_{\lambda\mu} \Gamma^\lambda_{\rho\nu} - \Gamma^\rho_{\lambda\nu} \Gamma^\lambda_{\rho\mu} \ , \tag{2.19.3}$$

and the $\boldsymbol{B}^{(3)}$ field [9—15] is proportional directly to the part of the Ricci tensor quadratic in the affine parameter. The latter is used to define the vector potential in curved space-time,

$$\Gamma^\kappa_{\lambda\mu} = \frac{e}{\hbar} M_\lambda A^\kappa_\mu \ , \tag{2.19.4}$$

where M_μ is a rotation generator [9,10] and where $eA^\kappa_\lambda / \hbar$ is an energy-momentum tensor of electromagnetism. If $\lambda = \kappa$ we obtain,

$$A_\mu := M_\lambda A^\lambda_\mu \ , \tag{2.19.5}$$

and the $G_{\mu\nu}$ tensor becomes,

$$G_{\mu\nu} = \partial_\mu A_\nu - \partial_\nu A_\mu + \frac{e}{\hbar} A^2 (M_\mu M_\nu - M_\nu M_\mu) \ . \tag{2.19.6}$$

Restricting attention to $\mu = 1$, $\nu = 2$, the $\boldsymbol{B}^{(3)}$ field in curved space-time is obtained as,

$$B_3 := \frac{e}{\hbar} A^2 (M_1 M_2 - M_2 M_1) \ , \tag{2.19.7}$$

or in the complex basis ((1), (2), (3)) [9—15],

$$\boldsymbol{B}^{(3)*} = -i\frac{e}{\hbar}\boldsymbol{A}^{(1)} \times \boldsymbol{A}^{(2)} . \tag{2.19.8}$$

Maxwellian electrodynamics is recovered if and only if the rotation generators commute, i.e., if the quadratic term is arbitrarily abandoned, so that

$$G_{\mu\nu} \rightarrow F_{\mu\nu} := \partial_\mu A_\nu - \partial_\nu A_\mu . \tag{2.19.9}$$

This means that Maxwellian electrodynamics is a linear approximation in which terms quadratic in A are missing. In the basis ((1), (2), (3)) the Maxwellian approximation means that the right hand side in the equation,

$$\boldsymbol{B}^{(1)} \times \boldsymbol{B}^{(2)} = iB^{(0)}\boldsymbol{B}^{(3)*} , \tag{2.19.10}$$

is zero. This is geometrically incorrect, and contradicts the experimental observation of $\boldsymbol{B}^{(1)} \times \boldsymbol{B}^{(2)}$ in magneto-optics [4—8]. Maxwellian electrodynamics is adequate therefore for low intensity light, but becomes internally inconsistent when magneto-optics is understood in terms of a self-consistent theory in curved space-time.

The $\boldsymbol{B}^{(3)}$ allows an understanding of electrodynamics and gravitation as being proportional to respectively the antisymmetric and symmetric components of the same Riemann tensor, i.e., in terms of curvilinear space-time geometry. If account is taken of $\boldsymbol{B}^{(3)}$ one can no longer logically adhere to a flat space-time for electromagnetism and a curved space-time for gravitation, and in magneto-optics, one is actually observing the spinning of space-time itself. Thus, as pointed out by Roy [11], $\boldsymbol{B}^{(3)}$ becomes the relict magnetic field in cosmology, responsible for circular polarization in the 2.7 K background radiation [19]. These are the essentials of electromagnetism in curved space-time rather than in flat space-time, and these ideas are being developed in detail elsewhere [20—22].

Acknowledgments

York University, Toronto, and the Indian Statistical Institute are thanked for visiting professorships, and the participants of the first Vigier Symposium are thanked for many interesting discussions.

References

[1] M. W. Evans, *Physica B* **182**, 227, 237 (1992); **183**, 103 (1993).
[2] M. W. Evans and S. Roy, *Found. Phys.*, submitted(1996); M. W. Evans, S. Roy, and S. Jeffers, *Il Nuovo Cim. B*, **110** 1473 (1995).
[3] M. W. Evans, *Physica B* **190**, 310 (1993); *Physica A* **214**, 605 (1995).
[4] reviewed by R. Zawodny in M. W. Evans and S. Kielich, eds., *Modern Nonlinear Optics,* Vols. 85(1), 85(2), 85(3) of *Advances in Chemical Physics,* I. Prigogine and S. A. Rice, eds., (Wiley Interscience, New York, 1993).
[5] S. Woźniak, M. W. Evans, and G. Wagnière, *Mol. Phys.* **75**, 81 (1992).
[6] P. S. Pershan, *Phys. Rev.* **130**, 919 (1963); J. P. van der Ziel, P. S. Pershan and L. D. Malmstrom, *Phys. Rev. Lett.* **15**, 190 (1965); *Phys. Rev.* **143**, 574 (1966).
[7] J. Deschamps, M. Fitaire, and M. Lagoutte, *Phys. Rev. Lett.* **25**, 1330 (1970); *Rev. Appl. Phys.* **7**, 155 (1972).
[8] W. Happer, *Rev. Mod. Phys.* **44**, 169 (1972).
[9] M. W. Evans and J.-P. Vigier, *The Enigmatic Photon, Vol. 1: The Field* $\boldsymbol{B}^{(3)}$ (Kluwer Academic, Dordrecht, 1994).
[10] M. W. Evans and J.-P. Vigier, *The Enigmatic Photon, Vol. 2: Non-Abelian Electrodynamics* (Kluwer Academic, Dordrecht, 1995).
[11] M. W. Evans, J.-P. Vigier, S. Roy, and S. Jeffers, *The Enigmatic Photon, Vol. 3: Theory and Practice of* $\boldsymbol{B}^{(3)}$ (Kluwer Academic, Dordrecht, 1996).
[12] M. W. Evans, *Found. Phys. Lett.* **7**, 67, 209, 379, 467, 577 (1994); **8**, 63, 83, 187, 363, 385 (1995).

[13] M. W. Evans, *Found. Phys.* **24**, 1519, 1671 (1994); **25**, 175, 383 (1995).
[14] M. W. Evans, in Ref. 4, Vol. 85(2); A. A. Hasanein and M. W. Evans, *The Photomagneton in Quantum Field Theory* (World Scientific, Singapore, 1994).
[15] M. W. Evans in A. van der Merwe and A. Garuccio, eds., *Waves and Particles in Light and Matter* (Plenum, New York, 1994).
[16] L. D. Landau and E. M. Lifshitz, *The Classical Theory of Fields*, 4th edn. (Pergamon, Oxford, 1975).
[17] C. W. Misner, K. S. Thorne, and J. A. Wheeler, *Gravitation* (W. H. Freeman and Co., San Francisco, 1973).
[18] S. K. Bose, *An Introduction to General Relativity* (Wiley Eastern Ltd., New Delhi, 1980).
[19] A. K. T. Assis and M. C. D. Neves, Apeiron, **2(3)**, 79 (1995).
[20] M. W. Evans, *Found. Phys. Lett.*, in press (1996).
[21] M. W. Evans, J.-P. Vigier, and S. Roy, *The Enigmatic Photon Vol. 4: New Directions* (Kluwer Academic, Dordrecht, 1998).
[22] M. W. Evans, *Found. Phys.*, submitted.

Paper 20

The Cyclic Structure of Vacuum Electromagnetism: Quantization and Derivation of Maxwell's Equations

Starting from the classical A cyclic equivalence principle of the new electrodynamics, the Faraday and Ampère laws are derived in quantized form, these being two of the Maxwell equations. The third A cyclic can be quantized self consistently using the same operators and de Broglie wavefunction. This method shows that: 1) if $\boldsymbol{B}^{(3)}$ =? $\mathbf{0}$ the Maxwell equations vanish; 2) there is no Faraday induction law for $\boldsymbol{B}^{(3)}$.

Key words: A cyclics, self-consistent quantization; Maxwell equations.

20.1 Introduction

The cyclic structure of the new electrodynamics based on the $\boldsymbol{B}^{(3)}$ field [1—7] gives an equivalence principle between the field and space-time, because, generally speaking, the structure of the field becomes the same as that of three dimensional space, described by the *O(3)* rotation group. In this Letter the second and third equations of the A cyclics [8] are quantized to give two of the vacuum Maxwell equations, the Faraday law and Ampère law with Maxwell's displacement current. The same method self-consistently quantizes the first equation of the A cyclics and shows that there is no Faraday induction law for $\boldsymbol{B}^{(3)}$. Consistently, no Faraday induction has been observed in a circularly polarized laser beam modulated inside an evacuated induction coil [1—4]. In this method, $\boldsymbol{A}^{(3)}$ quantizes to the $\hbar\partial/\partial Z$ operator and is not zero. If set to zero, all three A cyclics vanish, and with them the Maxwell equations. The Maxwell equations for $\boldsymbol{B}^{(1)} = \boldsymbol{B}^{(2)*}$ imply the existence of $\boldsymbol{B}^{(3)}$, and if the latter is set to zero arbitrarily, the Maxwell equations vanish. Finally the method allows direct quantization of the A cyclics to the Maxwell equations, which become equations of the quantum field theory. The method is therefore direct, simple, and easy to interpret.

20.2 Quantization of the Second and Third Equations

The A cyclic equivalence principle relies on the existence in the vacuum of a fully covariant four-vector whose four components are interrelated by [8]:

$$\boldsymbol{A}^{(1)} \times \boldsymbol{A}^{(2)} = iA^{(0)}\boldsymbol{A}^{(3)*}, \tag{2.20.1}$$

$$\boldsymbol{A}^{(2)} \times \boldsymbol{A}^{(3)} = iA^{(0)}\boldsymbol{A}^{(1)*}, \tag{2.20.2}$$

$$\boldsymbol{A}^{(3)} \times \boldsymbol{A}^{(1)} = iA^{(0)}\boldsymbol{A}^{(2)*}, \tag{2.20.3}$$

in the complex space basis ((1), (2), (3)) [1—4]. In this section, Eqs. (2.20.2) and (2.20.3) are quantized self-consistently to give two of the vacuum Maxwell equations, the Faraday Law and the Ampère law with

Maxwell's displacement current. Write Eq. (2.20.2) as the classical eigenvalue equation,

$$-\boldsymbol{A}^{(3)} \times \boldsymbol{A}^{(2)} = iA^{(0)}\boldsymbol{A}^{(2)}. \tag{2.20.4}$$

Use the minimal prescription in the form [8],

$$\boldsymbol{p}^{(3)} = ie\boldsymbol{A}^{(3)}, \qquad p^{(0)} = ieA^{(0)}, \tag{2.20.5}$$

and identify $\boldsymbol{A}^{(2)}$ with the classical eigenfunction $\Psi^{(2)}$. Here e is the elementary charge. This procedure results in the classical equation,

$$-\boldsymbol{p}^{(3)} \times \Psi^{(2)} = ip^{(0)}\Psi^{(2)}, \tag{2.20.6}$$

and the vector potential has taken on the dual role of operator and function in a classical eigenequation. Its ability to do this springs from the duality transform $\boldsymbol{A} \rightarrow i\boldsymbol{A}$ [9—12] in the complex three space ((1), (2), (3)). Therefore if $i\boldsymbol{A}$ is a polar vector multiplied by i, then $\boldsymbol{A}$ is an axial vector. The same duality transform takes the axial vector $\boldsymbol{B}$ to $i\boldsymbol{E}/c$, a polar vector multiplied by i. The fact that $\boldsymbol{A}$ is both polar and axial signifies that electromagnetism is chiral, with two enantiomeric forms — right and left circularly polarized [13]. Chirality in Dirac algebra becomes the eigenvalues of the γ_5 operator, playing the role of i in Pauli algebra [14]. This dual polar-axial nature of $\boldsymbol{A}$ allows it to be both an operator (polar vector) and function (axial vector).

The classical eigenvalue equation (2.20.6) is now quantized with the correspondence principle, whose operators $\boldsymbol{p}^{(3)} \rightarrow i\hbar\partial/\partial Z$ and $p^{(0)} \rightarrow -(i\hbar/c)(\partial/\partial t)$ act on a wavefunction in our complex three space. Let this wavefunction be [15],

$$\Psi^{(2)} = c\boldsymbol{B}^{(2)} - i\boldsymbol{E}^{(2)}, \tag{2.20.7}$$

as used by Majorana. Here c is the speed of light in vacuo, $\boldsymbol{B}$ is magnetic flux density and $\boldsymbol{E}$ is electric field strength. The function (2.20.7) includes the electromagnetic phase in the form of the scalar de Broglie wavefunction [16], and it is understood that the operators introduced by the correspondence principle operate on this. Therefore the operators $\boldsymbol{p}^{(3)}$ and $p^{(0)}$ are phase free, the function $\Psi^{(2)}$ is phase dependent. The quantum field equation derived in this way from the classical equation (2.20.6) is

$$\nabla \times \left(c\boldsymbol{B}^{(2)} - i\boldsymbol{E}^{(2)} \right) = \frac{i}{c}\frac{\partial}{\partial t}\left(c\boldsymbol{B}^{(2)} - i\boldsymbol{E}^{(2)} \right). \tag{2.20.8}$$

Compare real parts to give an equation of *quantized* field theory in the form of Ampère's law modified by Maxwell's vacuum displacement current,

$$\nabla \times \boldsymbol{B}^{(2)} = \frac{1}{c^2}\frac{\partial \boldsymbol{E}^{(2)}}{\partial t} . \tag{2.20.9}$$

Compare imaginary parts to give an equation of quantized field theory in the form of Faraday's law of induction,

$$\nabla \times \boldsymbol{E}^{(2)} = -\frac{\partial \boldsymbol{B}^{(2)}}{\partial t} . \tag{2.20.10}$$

Equations (2.20.9) and (2.20.10) are two of the four vacuum Maxwell equations, but have been derived through the correspondence principle and are therefore also equations of the quantum field theory. These take the same form as the classical Ampère-Maxwell and Faraday laws but are also equations of a novel, fully relativistic, quantum field theory.

Similarly, Eq. (2.20.3) quantizes to

$$\nabla \times \boldsymbol{B}^{(1)} = \frac{1}{c^2}\frac{\partial \boldsymbol{E}^{(1)}}{\partial t} , \tag{2.20.11}$$

$$\nabla \times \boldsymbol{E}^{(1)} = -\frac{\partial \boldsymbol{B}^{(1)}}{\partial t} . \qquad (2.20.12)$$

20.3 The d'Alembert Equation, Lorentz Condition and Acausal Energy Condition

The dual nature of the vector potential, once recognized, leads immediately to the d'Alembert equation, because A_μ is light-like. Therefore,

$$A_\mu A^\mu = 0, \qquad (2.20.13)$$

and taking the operator definition this becomes the d'Alembertian operating on a wavefunction in space-time, i.e.,

$$\partial_\mu \partial^\mu \psi_\nu = \Box \psi_\nu = 0. \qquad (2.20.14)$$

This is the quantized d'Alembert equation written for the four-vector ψ_ν. The latter in general has a space-like and time-like component. In this view A_μ must be a polar four-vector proportional to the generator of spacetime translations, and so the d'Alembert Eq. (2.20.14) is the first (mass) Casimir invariant of the Poincaré group [17]. The invariant is zero because we have assumed that c is the speed of light, and have taken photon mass to be zero.

If, in the condition $A_\mu A^\mu = 0$, we take the first A_μ as an operator through the correspondence principle, and interpret the second A^μ as a wavefunction ψ^μ, we obtain the quantized Lorentz condition for a massless particle,

$$\partial_\mu \psi^\mu = 0 . \qquad (2.20.15)$$

This is the orthogonality condition of the Poincaré group, which states that A_μ in operator form is orthogonal to A^μ in function form. The latter becomes the Pauli-Lubanski axial four-vector of the Poincaré group [18].

The condition $A_\mu A^\mu = 0$ interpreted as a condition on the wavefunction gives the acausal energy condition,

$$\psi_\mu \psi^\mu = 0, \tag{2.20.16}$$

which is the second (spin) invariant of the Poincaré group. Therefore we are dealing with a quantized particle with spin described by the three A cyclics (2.20.1—2.20.3). Evidently, this is the photon of the new relativistic quantum field theory developed here. The empirical evidence for the existence of this photon can be traced to the magneto-optical evidence for $\boldsymbol{B}^{(3)}$ in the inverse Faraday effect [1—4] and other effects. Without $\boldsymbol{B}^{(3)}$, this photon is undefined.

Finally, the energy condition (2.20.16) is the acausal solution suggested by Majorana [19]; Oppenheimer [20]; Dirac [21]; Wigner [22]; Gianetto [23] Ahluwalia and Ernst [24] and Chubykalo, Evans and Smirnov-Rueda [25]. It is longitudinal because the Pauli-Lubanski four-vector ψ_μ can be expressed in terms of the purely longitudinal [18,26],

$$\psi^\mu = cB^\mu + iE^\mu, \tag{2.20.17}$$

in the vacuum.

20.4 Self-Consistent Quantization of Equation (2.20.1)

The quantization of Eq. (2.20.1) occurs in a self-consistent way using the same operator interpretation of $iA^{(0)}$ and $iA^{(3)} = -iA^{(3)*}$. This gives the relativistic Schrödinger equation,

$$\frac{1}{c}\frac{\partial}{\partial t}\left(\frac{\partial}{\partial Z}\psi_0\right) = \left(\frac{e}{\hbar}A^{(0)}\right)^2 \psi_0 \, , \tag{2.20.18}$$

where ψ_0 is the scalar de Broglie wavefunction [17],

$$\psi_0 = \exp(i\phi) \, , \tag{2.20.19}$$

where $\phi = \omega t - \kappa Z$ is the electromagnetic phase. Here ω is the angular frequency at an instant t and κ the wavevector at point Z as usual. Using the vacuum minimal prescription [1—4],

$$eA^{(0)} = \hbar\kappa, \tag{2.20.20}$$

it is seen that Eq. (2.20.18) is self-consistent and consistent with the correspondence principle in the form (2.20.5). The method used to transform the second and third A cyclics into the Maxwell equations gives a fully consistent Schrödinger equation for the third cyclic. In this method $i\boldsymbol{A}^{(3)}$ is clearly not zero, and since $\boldsymbol{B}^{(3)} = \kappa\boldsymbol{A}^{(3)}$ [8], neither is $\boldsymbol{B}^{(3)}$. If we try to set $i\boldsymbol{A}^{(3)}$ to zero the del operator vanishes along with all three A cyclic equations. The Maxwell equations themselves vanish if we try $\boldsymbol{B}^{(3)} =? \mathbf{0}$. There is no vacuum Faraday induction law involving $\boldsymbol{B}^{(3)}$, because of the structure of Eq. (2.20.1), and this is again consistent with the experimental finding that there is no Faraday induction in a coil wound around a modulated monochromatic laser beam propagating in a vacuum [1—4]. The fundamental reason for this is that $\boldsymbol{B}^{(0)}$ is an unchanging property of one photon, i.e., $\hbar/e$ divided by the photon area.

20.5 Discussion

The duality transform $\boldsymbol{A} \rightarrow i\boldsymbol{A}$ in the vacuum shows that $\boldsymbol{A}$ can act as an operator and as a function. This transforms two of the A cyclic equations into two of the Maxwell equations in fully quantized form, producing a new quantum field theory for the photon, which acquires in the process three degrees of polarization. The first equation (2.20.1) of the A cyclics is quantized self-consistently. The structure of these equations shows that there is no Faraday induction law for $\boldsymbol{B}^{(3)}$, as observed experimentally. The explanation of magneto-optical phenomena [1—7] requires the use of the conjugate product $\boldsymbol{B}^{(1)} \times \boldsymbol{B}^{(2)}$; a product which demonstrates the existence of $iB^{(0)}\boldsymbol{B}^{(3)*}$ in the vacuum, and therefore of $\boldsymbol{B}^{(3)}$. Since $\boldsymbol{B}^{(3)}$ is $\kappa \boldsymbol{A}^{(3)}$, then an attempt to set $\boldsymbol{A}^{(3)}$ to zero removes the three equations of the A cyclics, and so removes the Maxwell equations themselves. Therefore the A and B cyclics become fundamental classical structures from which the Maxwell equations can be *derived* in quantized form using the correspondence principle.

There are clear differences between this new theory of electrodynamics and the received theory.

(1) The Maxwell equations are no longer the fundamental classical equations, they can be simultaneously derived and quantized from a more fundamental classical structure in which $\boldsymbol{B}$ and the rotational $\boldsymbol{A}$ are infinitesimal rotation generators of *O(3)*.

(2) The potential four-vector A_μ is fully covariant and has four non-zero components inter-related as in Eqs. (2.20.1) to (2.20.3). The older view allows a non-covariant A_μ such as the Coulomb gauge.

(3) The quantized d'Alembert equation becomes the first Casimir invariant of the Poincaré group; the quantized Lorentz condition becomes an orthogonality condition; and the quantized acausal energy condition becomes the second Casimir invariant. These results can be derived from the fact that A_μ plays the dual role of operator and function. Since $\boldsymbol{A}^{(3)}$ is directly proportional to $\boldsymbol{B}^{(3)}$ it is gauge invariant; a property which is consistent with the fact that the cross product $\boldsymbol{A}^{(1)} \times \boldsymbol{A}^{(2)}$ is gauge invariant [17] in the Poincaré group, but not in the U(1) group of the received view.

The most important and fundamental result of this analysis is that the Maxwell equations become derivative equations of a cyclical structure for electromagnetism in the vacuum. A similar result can be derived for the equations in the presence of sources (charges and currents).

Acknowledgments

Professor Erasmo Recami is thanked for sending several reprints describing the derivation by Majorana of the Maxwell equations in the form of a Dirac equation, producing in the process longitudinal solutions in the vacuum. The Majorana wavefunction is essential for the derivation of the Maxwell equations from the A (or B) cyclics.

References

[1] M. W. Evans and J.-P. Vigier, *The Enigmatic Photon, Vol. 1: The Field* $\boldsymbol{B}^{(3)}$ (Kluwer Academic, Dordrecht, 1994).

[2] M. W. Evans and J.-P. Vigier, *The Enigmatic Photon, Vol. 2: Non-Abelian Electrodynamics* (Kluwer Academic, Dordrecht, 1995).

[3] M. W. Evans, J.-P. Vigier, S. Roy, and S. Jeffers, *The Enigmatic Photon, Vol. 3: Theory and Practice of the* $\boldsymbol{B}^{(3)}$ *Field* (Kluwer Academic, Dordrecht, 1996).

[4] M. W. Evans, J.-P. Vigier, and, S. Roy, eds., *The Enigmatic Photon, Vol. 4: New Developments* (Kluwer Academic, Dordrecht, 1998), a collection of contributed papers.

[5] M. W. Evans and S. Kielich, eds., *Modern Nonlinear Optics,* Vol. 85(2) of *Advances in Chemical Physics,* I. Prigogine and S. A. Rice, eds. (Wiley Interscience, New York, 1993).

[6] A. A. Hasanein and M. W. Evans, *The Photomagneton in Quantum Field Theory* (World Scientific, Singapore, 1994).

[7] M. W. Evans, *The Photon's Magnetic Field* (World Scientific, Singapore, 1992).

[8] M. W. Evans, *Found. Phys. Lett.* **8**, 63, 83, 187, 363, 385 (1995).

[9] R. Mignani, E. Recami and M. Baldo, *Lett. Nuovo Cim.* **11**, 568 (1974).

[10] E. Giannetto, *Lett. Nuovo Cim.* **44**, 140, 145 (1985).

[11] R. Mignani and E. Recami, *Nuovo Cim.* **30A**, 533 (1975).

[12] R. Mignani and E. Recami, *Phys. Lett.* **62B**, 41 (1976).

[13] L. D. Landau and E. M. Lifshitz, *The Classical Theory of Fields*, 4th edn. (Pergamon, Oxford, 1975).
[14] M. A. Defaria-Rosa, E. Recami, and W. A. Rodrigues Jr., *Phys. Lett.* **173B**, 233 (1986).
[15] E. Majorana, Quaderno 2, Folio 101/1 of Scientific Manuscripts, circa 1928 to 1932, Domus Galilaeana, Pisa, reproduced in Ref. 9.
16] L. de Broglie, *C. R. Acad. Sci.* **177**, 507 (1923); *Phil. Mag.* **47**, 446 (1924); *Ann. Phys. (Paris)* **3**, 22 (1925).
[17] L. H. Ryder, *Quantum Field Theory*, 2nd edn. (Cambridge University Press, Cambridge, 1987).
[18] M. W. Evans, *Physica A* **214**, 605 (1995); see also Chap. 11 of Ref. 1.
[19] E. Majorana, *Nuovo Cim.* **9**, 335 (1932).
[20] J. R. Oppenheimer, *Phys. Rev.* **38**, 725 (1931); ibid., in B. S. de Witt, ed., *Lectures on Electrodynamics* (New York, 1970).
[21] P. A. M. Dirac, *Phys. Rev.* **74**, 817 (1948).
[22] E. P. Wigner, *Ann. Math.* **40**, 149 (1939).
[23] E. Giannetto, *Lett. Nuov. Cim.* **44**, 140 (1985).
[24] D. V. Ahluwalia and D. J. Ernst, *Mod. Phys. Lett.* **7A**, 1967 (1992).
[25] A. E. Chubykalo, M. W. Evans, and R. Smirnov-Rueda, *Found. Phys. Lett.* **10,** 93 (1997).
[26] V. V. Dvoeglazov, *Phys. Rev. D* in press (1996); *To the Claimed* Longitudity of the Antisymmetric Tensor Field after Quantization

Index

Contents, Volume 1

Appendices

Contents, Volume 2

Contents, Volume 3

Contents, Volume 4

23. W.T. Grandy, Jr.: *Foundations of Statistical Mechanics.*
Vol. II: *Nonequilibrium Phenomena.* 1988 ISBN 90-277-2649-3
24. E.I. Bitsakis and C.A. Nicolaides (eds.): *The Concept of Probability.* Proceedings of the Delphi Conference (Delphi, Greece, 1987). 1989 ISBN 90-277-2679-5
25. A. van der Merwe, F. Selleri and G. Tarozzi (eds.): *Microphysical Reality and Quantum Formalism, Vol. 1.* Proceedings of the International Conference (Urbino, Italy, 1985). 1988 ISBN 90-277-2683-3
26. A. van der Merwe, F. Selleri and G. Tarozzi (eds.): *Microphysical Reality and Quantum Formalism, Vol. 2.* Proceedings of the International Conference (Urbino, Italy, 1985). 1988 ISBN 90-277-2684-1
27. I.D. Novikov and V.P. Frolov: *Physics of Black Holes.* 1989 ISBN 90-277-2685-X
28. G. Tarozzi and A. van der Merwe (eds.): *The Nature of Quantum Paradoxes.* Italian Studies in the Foundations and Philosophy of Modern Physics. 1988 ISBN 90-277-2703-1
29. B.R. Iyer, N. Mukunda and C.V. Vishveshwara (eds.): *Gravitation, Gauge Theories and the Early Universe.* 1989 ISBN 90-277-2710-4
30. H. Mark and L. Wood (eds.): *Energy in Physics, War and Peace.* A Festschrift celebrating Edward Teller's 80th Birthday. 1988 ISBN 90-277-2775-9
31. G.J. Erickson and C.R. Smith (eds.): *Maximum-Entropy and Bayesian Methods in Science and Engineering.*
Vol. I: *Foundations.* 1988 ISBN 90-277-2793-7
32. G.J. Erickson and C.R. Smith (eds.): *Maximum-Entropy and Bayesian Methods in Science and Engineering.*
Vol. II: *Applications.* 1988 ISBN 90-277-2794-5
33. M.E. Noz and Y.S. Kim (eds.): *Special Relativity and Quantum Theory.* A Collection of Papers on the Poincaré Group. 1988 ISBN 90-277-2799-6
34. I.Yu. Kobzarev and Yu.I. Manin: *Elementary Particles. Mathematics, Physics and Philosophy.* 1989 ISBN 0-7923-0098-X
35. F. Selleri: *Quantum Paradoxes and Physical Reality.* 1990 ISBN 0-7923-0253-2
36. J. Skilling (ed.): *Maximum-Entropy and Bayesian Methods.* Proceedings of the 8th International Workshop (Cambridge, UK, 1988). 1989 ISBN 0-7923-0224-9
37. M. Kafatos (ed.): *Bell's Theorem, Quantum Theory and Conceptions of the Universe.* 1989 ISBN 0-7923-0496-9
38. Yu.A. Izyumov and V.N. Syromyatnikov: *Phase Transitions and Crystal Symmetry.* 1990 ISBN 0-7923-0542-6
39. P.F. Fougère (ed.): *Maximum-Entropy and Bayesian Methods.* Proceedings of the 9th International Workshop (Dartmouth, Massachusetts, USA, 1989). 1990 ISBN 0-7923-0928-6
40. L. de Broglie: *Heisenberg's Uncertainties and the Probabilistic Interpretation of Wave Mechanics.* With Critical Notes of the Author. 1990 ISBN 0-7923-0929-4
41. W.T. Grandy, Jr.: *Relativistic Quantum Mechanics of Leptons and Fields.* 1991 ISBN 0-7923-1049-7
42. Yu.L. Klimontovich: *Turbulent Motion and the Structure of Chaos.* A New Approach to the Statistical Theory of Open Systems. 1991 ISBN 0-7923-1114-0
43. W.T. Grandy, Jr. and L.H. Schick (eds.): *Maximum-Entropy and Bayesian Methods.* Proceedings of the 10th International Workshop (Laramie, Wyoming, USA, 1990). 1991 ISBN 0-7923-1140-X

44. P.Pták and S. Pulmannová: *Orthomodular Structures as Quantum Logics.* Intrinsic Properties, State Space and Probabilistic Topics. 1991 ISBN 0-7923-1207-4
45. D. Hestenes and A. Weingartshofer (eds.): *The Electron.* New Theory and Experiment. 1991 ISBN 0-7923-1356-9
46. P.P.J.M. Schram: *Kinetic Theory of Gases and Plasmas.* 1991 ISBN 0-7923-1392-5
47. A. Micali, R. Boudet and J. Helmstetter (eds.): *Clifford Algebras and their Applications in Mathematical Physics.* 1992 ISBN 0-7923-1623-1
48. E. Prugovečki: *Quantum Geometry.* A Framework for Quantum General Relativity. 1992 ISBN 0-7923-1640-1
49. M.H. Mac Gregor: *The Enigmatic Electron.* 1992 ISBN 0-7923-1982-6
50. C.R. Smith, G.J. Erickson and P.O. Neudorfer (eds.): *Maximum Entropy and Bayesian Methods.* Proceedings of the 11th International Workshop (Seattle, 1991). 1993 ISBN 0-7923-2031-X
51. D.J. Hoekzema: *The Quantum Labyrinth.* 1993 ISBN 0-7923-2066-2
52. Z. Oziewicz, B. Jancewicz and A. Borowiec (eds.): *Spinors, Twistors, Clifford Algebras and Quantum Deformations.* Proceedings of the Second Max Born Symposium (Wrocław, Poland, 1992). 1993 ISBN 0-7923-2251-7
53. A. Mohammad-Djafari and G. Demoment (eds.): *Maximum Entropy and Bayesian Methods.* Proceedings of the 12th International Workshop (Paris, France, 1992). 1993 ISBN 0-7923-2280-0
54. M. Riesz: *Clifford Numbers and Spinors* with Riesz' Private Lectures to E. Folke Bolinder and a Historical Review by Pertti Lounesto. E.F. Bolinder and P. Lounesto (eds.). 1993 ISBN 0-7923-2299-1
55. F. Brackx, R. Delanghe and H. Serras (eds.): *Clifford Algebras and their Applications in Mathematical Physics.* Proceedings of the Third Conference (Deinze, 1993) 1993 ISBN 0-7923-2347-5
56. J.R. Fanchi: *Parametrized Relativistic Quantum Theory.* 1993 ISBN 0-7923-2376-9
57. A. Peres: *Quantum Theory: Concepts and Methods.* 1993 ISBN 0-7923-2549-4
58. P.L. Antonelli, R.S. Ingarden and M. Matsumoto: *The Theory of Sprays and Finsler Spaces with Applications in Physics and Biology.* 1993 ISBN 0-7923-2577-X
59. R. Miron and M. Anastasiei: *The Geometry of Lagrange Spaces: Theory and Applications.* 1994 ISBN 0-7923-2591-5
60. G. Adomian: *Solving Frontier Problems of Physics: The Decomposition Method.* 1994 ISBN 0-7923-2644-X
61 B.S. Kerner and V.V. Osipov: *Autosolitons.* A New Approach to Problems of Self-Organization and Turbulence. 1994 ISBN 0-7923-2816-7
62. G.R. Heidbreder (ed.): *Maximum Entropy and Bayesian Methods.* Proceedings of the 13th International Workshop (Santa Barbara, USA, 1993) 1996 ISBN 0-7923-2851-5
63. J. Peřina, Z. Hradil and B. Jurčo: *Quantum Optics and Fundamentals of Physics.* 1994 ISBN 0-7923-3000-5
64. M. Evans and J.-P. Vigier: *The Enigmatic Photon.* Volume 1: The Field $\boldsymbol{B}^{(3)}$. 1994 ISBN 0-7923-3049-8
65. C.K. Raju: *Time: Towards a Constistent Theory.* 1994 ISBN 0-7923-3103-6
66. A.K.T. Assis: *Weber's Electrodynamics.* 1994 ISBN 0-7923-3137-0
67. Yu. L. Klimontovich: *Statistical Theory of Open Systems.* Volume 1: A Unified Approach to Kinetic Description of Processes in Active Systems. 1995 ISBN 0-7923-3199-0; Pb: ISBN 0-7923-3242-3

Fundamental Theories of Physics

68. M. Evans and J.-P. Vigier: *The Enigmatic Photon.* Volume 2: Non-Abelian Electrodynamics. 1995 ISBN 0-7923-3288-1
69. G. Esposito: *Complex General Relativity.* 1995 ISBN 0-7923-3340-3
70. J. Skilling and S. Sibisi (eds.): *Maximum Entropy and Bayesian Methods.* Proceedings of the Fourteenth International Workshop on Maximum Entropy and Bayesian Methods. 1996 ISBN 0-7923-3452-3
71. C. Garola and A. Rossi (eds.): *The Foundations of Quantum Mechanics – Historical Analysis and Open Questions.* 1995 ISBN 0-7923-3480-9
72. A. Peres: *Quantum Theory: Concepts and Methods.* 1995 (see for hardback edition, Vol. 57) ISBN Pb 0-7923-3632-1
73. M. Ferrero and A. van der Merwe (eds.): *Fundamental Problems in Quantum Physics.* 1995 ISBN 0-7923-3670-4
74. F.E. Schroeck, Jr.: *Quantum Mechanics on Phase Space.* 1996 ISBN 0-7923-3794-8
75. L. de la Peña and A.M. Cetto: *The Quantum Dice.* An Introduction to Stochastic Electrodynamics. 1996 ISBN 0-7923-3818-9
76. P.L. Antonelli and R. Miron (eds.): *Lagrange and Finsler Geometry.* Applications to Physics and Biology. 1996 ISBN 0-7923-3873-1
77. M.W. Evans, J.-P. Vigier, S. Roy and S. Jeffers: *The Enigmatic Photon.* Volume 3: Theory and Practice of the $\boldsymbol{B}^{(3)}$ Field. 1996 ISBN 0-7923-4044-2
78. W.G.V. Rosser: *Interpretation of Classical Electromagnetism.* 1996 ISBN 0-7923-4187-2
79. K.M. Hanson and R.N. Silver (eds.): *Maximum Entropy and Bayesian Methods.* 1996 ISBN 0-7923-4311-5
80. S. Jeffers, S. Roy, J.-P. Vigier and G. Hunter (eds.): *The Present Status of the Quantum Theory of Light.* Proceedings of a Symposium in Honour of Jean-Pierre Vigier. 1997 ISBN 0-7923-4337-9
81. M. Ferrero and A. van der Merwe (eds.): *New Developments on Fundamental Problems in Quantum Physics.* 1997 ISBN 0-7923-4374-3
82. R. Miron: *The Geometry of Higher-Order Lagrange Spaces.* Applications to Mechanics and Physics. 1997 ISBN 0-7923-4393-X
83. T. Hakioğlu and A.S. Shumovsky (eds.): *Quantum Optics and the Spectroscopy of Solids.* Concepts and Advances. 1997 ISBN 0-7923-4414-6
84. A. Sitenko and V. Tartakovskii: *Theory of Nucleus.* Nuclear Structure and Nuclear Interaction. 1997 ISBN 0-7923-4423-5
85. G. Esposito, A.Yu. Kamenshchik and G. Pollifrone: *Euclidean Quantum Gravity on Manifolds with Boundary.* 1997 ISBN 0-7923-4472-3
86. R.S. Ingarden, A. Kossakowski and M. Ohya: *Information Dynamics and Open Systems.* Classical and Quantum Approach. 1997 ISBN 0-7923-4473-1
87. K. Nakamura: *Quantum versus Chaos.* Questions Emerging from Mesoscopic Cosmos. 1997 ISBN 0-7923-4557-6
88. B.R. Iyer and C.V. Vishveshwara (eds.): *Geometry, Fields and Cosmology.* Techniques and Applications. 1997 ISBN 0-7923-4725-0
89. G.A. Martynov: *Classical Statistical Mechanics.* 1997 ISBN 0-7923-4774-9
90. M.W. Evans, J.-P. Vigier, S. Roy and G. Hunter (eds.): *The Enigmatic Photon.* Volume 4: New Directions. 1998 ISBN 0-7923-4826-5
91. M. Rédei: *Quantum Logic in Algebraic Approach.* 1998 ISBN 0-7923-4903-2
92. S. Roy: *Statistical Geometry and Applications to Microphysics and Cosmology.* 1998 ISBN 0-7923-4907-5

Fundamental Theories of Physics

93. B.C. Eu: *Nonequilibrium Statistical Mechanics*. Ensembled Method. 1998 ISBN 0-7923-4980-6
94. V. Dietrich, K. Habetha and G. Jank (eds.): *Clifford Algebras and Their Application in Mathematical Physics*. Aachen 1996. 1998 ISBN 0-7923-5037-5
95. *Not yet known*
96. V.P. Frolov and I.D. Novikov: *Black Hole Physics*. Basic Concepts and New Developments. 1998 ISBN 0-7923-5145-2; PB 0-7923-5146
97. G. Hunter, S. Jeffers and J-P. Vigier (eds.): *Causality and Locality in Modern Physics*. 1998 ISBN 0-7923-5227-0
98. G.J. Erickson, J.T. Rychert and C.R. Smith (eds.): *Maximum Entropy and Bayesian Methods*. 1998 ISBN 0-7923-5047-2
99. D. Hestenes: *New Foundations for Classical Mechanics (Second Edition)*. 1999 ISBN 0-7923-5302-1; PB ISBN 0-7923-5514-8
100. B.R. Iyer and B. Bhawal: *Black Holes, Gravitational Radiation and the Universe*. Essays in Honor of C. V. Vishveshwara. 1999 ISBN 0-7923-5308-0
101. P.L. Antonelli and T.J. Zastawniak: *Fundamentals of Finslerian Diffusion with Applications*. 1999 ISBN 0-7923-5511-3

KLUWER ACADEMIC PUBLISHERS – DORDRECHT / BOSTON / LONDON